Fluidic Systems Design

FLUIDIC SYSTEMS DESIGN

CHARLES A. BELSTERLING

WILEY-INTERSCIENCE, a Division of John Wiley & Sons, Inc.
New York · London · Sydney · Toronto

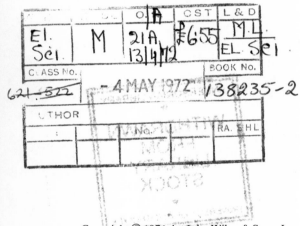

Copyright © 1971, by John Wiley & Sons, Inc.

All rights reserved. Published simultaneously in Canada.

No part of this book may be reproduced by any means, nor transmitted, nor translated into a machine language without the written permission of the publisher.

Library of Congress Catalogue Card Number: 79-148500

ISBN 0-471-06415-7

Printed in the United States of America.

10 9 8 7 6 5 4 3 2 1

Preface

Fluidic Systems Design was conceived and prepared to satisfy a widely recognized need for analytical methods for designing fluidic systems. Its primary purpose is to provide the control engineer with a unified set of analytical tools for the straightforward design of systems using fluidic devices. A second aim is to supply a universally acceptable vocabulary so that the control engineer, the fluidic device manufacturer, and the user's project engineer can communicate in a common language. In short, the intent is to bring together and make known the available techniques for describing fluidic components and for designing fluidic systems.

Most of the design methods described in this book are the result of a research program conducted since February 1963. At that time I was employed by the Franklin Institute Research Laboratories and was involved in the development of high-performance electric, hydraulic, and pneumatic control systems. We had been aware of the fluidics technology since its introduction by the Diamond Ordnance Fuze Laboratories (now Harry Diamond Laboratories) in 1959. At least one of our programs could have benefited from the unique features of fluidic devices. Therefore we began immediately to study their characteristics.

It did not take very long to discover that, first, the device manufacturers could not describe their components in terms familiar to control systems engineers and, second, that there were no proven methods for predicting the behavior of fluidic components connected together into control systems. I immediately prepared a detailed plan for a research program to correct this situation and was rewarded by financial support on a Laboratory-sponsored project. By October, I and my associate Kacheung Tsui had made enough progress to convince Harry Diamond Laboratories of the potential of the work. They provided the financial support and technical direction for the second phase.

After joining Giannini Controls Corporation (now Conrac Corporation) in July 1964, I was fortunate to enlist further support from the U.S. Army Aviation Materiel Laboratories. We completed the work on analog systems and published *Fluidic Systems Design Manual* in May 1967.

By January 1968 we had formed Cutler Controls, and I began to expand the design manual to cover digital as well as analog devices. I also investigated rather closely the work of other people in this field to establish the timeliness and permanence of a book on the subject. As a result I considered it necessary to include a chapter placing this book in its proper perspective and a chapter briefly touching on other techniques under development to accomplish the same objectives.

To communicate the contents of this book to its readers with the least amount of excess material, I have assumed a level of education equal to a bachelor of science degree in electrical or mechanical engineering. In addition I assume that the reader has a working knowledge of the following topics:

1. Electrical circuit analysis
2. Transform calculus
3. Feedback control systems
4. Matrix algebra

This book covers the design of control systems using analog and digital fluidic devices. The methods presented are applicable to all kinds of fluidic components: sensors, amplifiers, logic elements, networks, and actuators. They are applicable to all kinds of fluids: gases, compressible liquids (such as hydraulic fluids), and incompressible liquids (such as water). They are applicable to all kinds of systems: aircraft, spacecraft, ships, land vehicles, automatic machines, computers, and tracking devices. They are applicable to a broad spectrum of performance situations: static, large signal, dynamic, and small-signal at frequencies up to a level where the physical dimensions of the circuit approach the wavelength of the signal. They are applicable to practical situations in which second-order effects of the behavior of fluids can be safely neglected. In other words, they cover most normal cases.

It should be emphasized that, although the techniques described in this book are applicable to digital as well as analog systems, in many cases they will not be required for digital systems design. This is because more development attention has been devoted to digital devices; hence, the manufacturer is able to supply a line of components designed so that the characteristics are already matched for easy coupling.

Fluidic Systems Design is divided into seven major sections. Chapter 1 is a review of the fluidics technology, intended to place the book in its proper perspective. Chapters 2 and 3 introduce the subject of the book in general terms. Chapter 4 covers in detail the description of static characteristics and dynamic characteristics of fluidic components.

Chapters 5 and 6 describe the test methods and instrumentation required to measure component and system characteristics. Chapters 7 through 10 cover systems synthesis, both static large-signal, and dynamic small-signal summarizing with a design check list. Finally, the Appendix contains appropriate standards for the fluidics technology and an illustrative design problem.

To summarize, *Fluidic Systems Design* is my personal contribution to organizing and promoting useful fluidic systems design methods. These design methods have been proved valid for most *practical* cases, and the sheer insight they provide helps to remove the shroud of mystery that has surrounded the fluidics technology and inhibited its normal progress. As a result I draw the following conclusions concerning the status of the fluidics systems design technology:

1. No "breakthroughs" are required—the necessary analytical tools are on hand to design sophisticated fluidic systems.
2. The lack of a thorough mathematical description of the behavior of fluids in fluidic devices is not a handicap.
3. Graphical characteristics are the mainstay of current systems design.
4. Physical models, such as electric equivalent circuits, are currently sufficient for analyzing the dynamic characteristics of high-performance fluidic systems.
5. Mathematical models based on matrices will be the subject of most future effort, resulting in a comprehensive description of component characteristics including all cross-coupling terms.

I gratefully acknowledge the support and aid of the following persons: C. W. Hargens and Kacheung Tsui of the Franklin Institute Research Laboratories; Joseph Kirshner and Silas Katz of Harry Diamond Laboratories; Melvin Zisfein and Norris Barr of the Conrac Corporation; George Fosdick of the U.S. Army AVLABS; and Milton C. Stone of Cutler Controls, Inc.

CHARLES A. BELSTERLING

King of Prussia, Pennsylvania
October 1970

Contents

Chapter 1 History of the Development of Fluidic Systems Design Techniques — **1**

    1.1 Introduction and Scope,   1
    1.2 Definition of the Problems,   2
    1.3 Review of the Electron Tube Technology,   4
    1.4 Review of the Pneumatics Technology,   7
    1.5 Review of the Hydraulics Technology,   9
    1.6 Review of the Transistor Technology,   11
    1.7 Review of Nonlinear Circuit Analysis,   13
    1.8 Summary of the Review of Other Technologies,   13
    1.9 Review of the Fluidics Technology,   14

Chapter 2 The Fluidic Systems Design Process — **31**

    2.1 The General Approach,   31
    2.2 Definition of System Requirements,   32
    2.3 Component Requirements,   33
    2.4 Description of Components,   33
    2.5 Component Selection,   39
    2.6 Calculation of Performance,   40
    2.7 Parameter Studies,   43
    2.8 Summary of the Design Process,   43

Chapter 3 Operating Principles of Fluidic Devices — **44**

    3.1 Analog Amplifiers,   44
            Vented Jet-Interaction Amplifier,   44
            Closed Jet-Interaction Amplifier,   46
            Vortex Amplifier,   48
            Boundary-Layer-Control Amplifier,   50
            Impact Modulator,   52
    3.2 Analog Sensors,   53
            Back-Pressure Nozzle,   53

x　Contents

		Interruptable Jet,	54
		Bubbler Tubes,	55
		Converging Jet Sensor,	56
		Diverging Jet Sensor,	57
		Vortex Proximity Sensor,	57
		Vortex Rate Sensor,	58
	3.3	Digital Amplifiers,	59
		Wall Attachment Amplifier,	59
		Turbulence Amplifier,	61
		Axisymmetric Focused-Jet Amplifier,	62
		Passive Logic Devices,	64
	3.4	Digital Sensors,	65
		Limit Valves,	65
		Interrupted Laminar Jet,	66
		Diverging Jet Sensor,	67
		Acoustic Beam Sensors,	68
Chapter 4	**Component Description**		**69**
	4.1	Graphical Characteristics,	69
		Analog Amplifiers,	70
		Digital Amplifiers,	73
		Normalization of Characteristic Curves,	75
	4.2	Equivalent Electric Circuits,	76
		Analog Amplifiers,	76
		Digital Amplifiers,	77
	4.3	Important Physical Quantities,	77
		Line Lengths,	77
		Jet Path Lengths,	78
		Effective Areas,	78
		Effective Volumes,	78
		Power Jet Velocity,	78
		Qualities of the Operating fluid,	79
	4.4	Performance Parameters,	79
		Output Resistance R_o,	79
		Pressure Gain G_p,	79
		Pressure Amplification Factor K_p,	80
		Flow Gain G_f,	81
		Sensitivity Factor K,	81
		Input Resistance R_c,	81
		Equivalent Capacitance C,	82
		Equivalent Inductance L,	82
		Time Delay t_d,	83

			Contents	xi

		Pressure Recovery Factor R_p,	83
		Signal-to-Noise Ratio S/N,	83

Chapter 5 Test Methods and Instrumentation — 85

- 5.1 Static, — 85
  - Analog Amplifiers, — 85
  - Digital Amplifiers, — 88
  - Vortex Rate Sensor, — 88
  - Impedances, — 90
- 5.2 Dynamic, — 92
  - Analog Amplifiers, — 92
  - Frequency Response Testing, — 95
  - Measuring Total Time Delay, — 95
  - Digital Devices, — 97
  - Sensors, — 98
  - Impedances, — 99
- 5.3 Special Test Equipment, — 101
  - Pneumatic Pressure Signal Generator, — 101
  - Pressure Transducers, — 101
  - Flow Transducers, — 101
  - XY Oscilloscope, — 102
  - Automated Equipment, — 103

Chapter 6 Graphical Characteristics of Typical Fluidic Devices — 104

- 6.1 Turbulent (Nonlinear) Restrictors, — 104
- 6.2 Laminar (Linear) Restrictors, — 104
- 6.3 Analog Amplifiers, — 106
  - Vented Jet-Interaction Amplifier Output Characteristics, — 106
  - Vented Jet-Interaction Amplifier Input Characteristics, — 106
  - Vented Jet-Interaction Amplifier Transfer Characteristics, — 106
  - Closed Jet-Interaction Amplifier Output Characteristics, — 106
  - Vented Elbow Amplifier Output Characteristics, — 108
- 6.4 Digital Amplifiers, — 109
  - Vented Wall-Attachment Flip-Flop Output Characteristics, — 109
  - Vented Wall-Attachment Flip-Flop Input Characteristics, — 110
  - Turbulence Amplifier Output Characteristics, — 110
  - Turbulence Amplifier Input Characteristics, — 110

xii Contents

      6.5 Sensors, 111
            Back-Pressure Nozzle Output Characteristics, 111
            Interruptable Jet Output Characteristics, 111
            Vortex Rate Sensor Output Characteristics, 111
      6.6 Actuators, 113
            Input Characteristics of Rectilinear and Rotary Actuators, 113
            Input Characteristics of Fluid Motors, 114

Chapter 7   **Large-Signal Performance Analysis**   **115**
      7.1 The Load Line, 115
            Load Line Design Procedure, 119
      7.2 Calculation of the Transfer (Gain) Curve, 120
            Transfer Curve Design Procedure, 120
      7.3 Static Matching of Cascaded Fluidic Components, 122
            Objectives, 122
            Matching a Vortex Rate Sensor and a Differential Amplifier, 122
            Providing Proper Gains, 123
            Matching Operating Bias Points, 125
            Matching Operating Ranges, 125
            Matching a Digital Amplifier with Three Parallel Flip-Flops, 127
            Matching Operating Bias Points, 129
            Matching Operating Ranges, 130
            Procedure For Matching, 132

Chapter 8   **Equivalent Circuits for Typical Fluidic Devices**   **133**
      8.1 Analog Fluidic Amplifiers, 133
            Equivalent Circuit for a Vented Jet-Interaction Amplifier, 136
            Equivalent Circuit for a Closed Jet-Interaction Amplifier, 137
            Equivalent Circuit for a Vented Elbow Amplifier, 138
      8.2 Digital Fluidic Amplifiers, 139
            Equivalent Circuit for a Vented Wall-Attachment Amplifier, 139
      8.3 Fluidic Sensors, 140
            Equivalent Circuit for a Vortex Rate Sensor, 140
      8.4 Actuators, 141
            Equivalent Circuit for a Piston-Type Actuator, 141
            Equivalent Circuit for Piston-Type and Vane-Type Motors, 141

Chapter 9	**Small-Signal Performance Analysis**		**143**
	9.1	Derivation of the Transfer Function for Cascaded Fluidic Components,	143
	9.2	Calculation of Equivalent Circuit Parameters,	147
		Vortex Rate Sensor,	147
		Amplifier,	148
	9.3	Calculation of Frequency Response,	150
		Substitution of the Variable,	150
		Calculation of Linear Response,	151
		Calculation of the Response to Time Delay,	151
		The Total Frequency Response,	153
Chapter 10	**Detailed Systems Design Procedure**		**154**
	10.1	Required Information,	154
	10.2	Step-By-Step Design,	154
	10.3	Design Check List,	156
Appendix A	**Applicable Standards**		**157**
	A.1	Terminology,	157
		General,	157
		Amplifiers,	158
		Sensors,	162
		Transducers,	162
		Actuators,	162
		Displays,	162
		Logic Devices,	162
		Circuit Elements,	162
	A.2	Nomenclature and Units,	163
		Basic Quantities,	163
		General Subscripts,	164
	A.3	Graphical Symbology,	164
		General Conventions,	165
		Analog Fluidic Devices,	166
		Bistable Digital Devices,	169
		Monostable Digital Devices,	171
		Passive Digital Devices,	173
		Fluidic Impedances,	173
Appendix B	**Illustrative Example of V/STOL Control System Design**		**175**
	B.1	Description of the UH-1B Yaw Damper System,	175
	B.2	Required Information	175
		Performance Specifications for the System,	175

		Characteristics of Available Supply, Signal Sources, and Driven Load,	175
		Static Characteristics of Available Components,	176
		Internal Dimensional Characteristics of Available Components,	177
	B.3	Step-By-Step Design,	181
		Rate Sensor and First Stage Amplifier,	183
		First and Second Stage Amplifiers,	187
		Second and Third Stage Amplifiers,	191
		Third and Fourth Stage Amplifiers,	192
		Fifth Stage Amplifier with Servo Load,	194
		Fourth and Fifth Stage Amplifiers,	195
		Complete Fluidic System,	197

References 225

Index 227

Fluidic Systems Design

1
History of the Development of Fluidic Systems Design Techniques

1.1 INTRODUCTION AND SCOPE

To place this book in its proper perspective the first chapter is devoted to a review and evaluation of the evolution of the development of techniques for describing fluidic components and synthesizing fluidic systems.

The review is presented as follows:

1. Definition of the problem.
2. Description of solutions used in other technologies for solving similar problems.
3. Review of the history of development of analytical techniques for fluidic systems.
4. Definition of the major milestones in this development.

An extensive bibliography is included at the end of this chapter.

At the outset it is important to define the scope of this review. The useful frequency spectrum for fluidic systems can be divided into three distinct bands as shown in Figure 1.1. These bands can be described as follows:

Band 1: Static and low frequencies where frequency-sensitive characteristics (time-dependent variables) are not important.

Band 2: Higher frequencies where the dynamic (time-dependent) characteristics are important and can be represented by lumped parameters.

2 History of Fluidic Systems Design Techniques

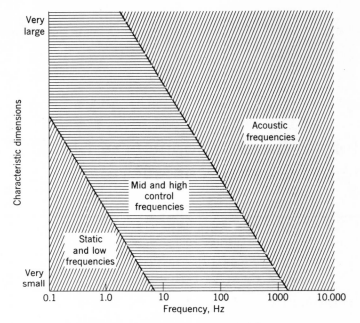

Figure 1.1 The analytical spectrum for fluidic systems.

Band 3: Acoustic frequencies where the wave length of the signal approaches the dimensions of the circuit and the characteristics must be represented as a distributed parameter system.

The major portions of the practical spectrum for fluidic device application—bands 1 and 2—will be considered in this review. This scope includes frequencies up to the kilohertz range if the characteristic dimensions of the devices and circuits are relatively small, but only into the 100 Hz range for larger devices. These ranges are adequate to cover most practical situations, in fact they easily encompass the performance requirements for most real applications.

1.2 DEFINITION OF THE PROBLEMS

In any system design the effect of one interconnected component upon another must be taken into account. The simplest illustration of this is the change in the behavior of any given component because of a load.

These effects must be considered for all components; electronic, mechanical, hydraulic, acoustic, pneumatic, or fluidic.

It must also be recognized that systems are *collections* of components. The number of possible combinations of components is very large or at least too large to be preplanned. Methods for analyzing and synthesizing systems must describe each isolated component and its relationship with any other device in the system. This concept in its simplest form is commonly known as the "black box" approach. However, it embraces far more than is usually associated with the common term. For example, a given component (or element) can be described in various ways:

1. A set of coupled equations.
2. A set of graphical characteristics.
3. An analogous circuit or model.

As stated by F. T. Brown[59] the optimum analytical method should include all significant information (be complete and accurate) as well as provide physical insight (be readily understood and referrable to known phenomena). Unfortunately the types of component descriptions listed above do not meet these criteria. That is, a set of coupled equations can be as accurate and complete as one cares to make it, by including nonlinearities and all possible coupling terms. However, little physical insight is associated with such a complex array. On the other hand, analogous circuits and models are usually limited to small perturbations and often involve gross oversimplifications because the elements are assumed to be linear. But analogs provide a tremendous "feel" for the behavior of the device. The graphical approach lies somewhere between the other two descriptive methods. This procedure can account for most nonlinear terms but fails to provide some of the information necessary for an intuitive appreciation of the behavior of the device.

The fluidic technology is certainly not the first to be faced with the need for developing analytical methods for nonlinear component analysis and systems synthesis. For example, the vacuum tube disciplines demanded entirely new approaches to electric circuit analysis, and new methods had to be developed when the dynamic behavior of pneumatic and hydraulic systems became important.

Fluidic technology also is not the first to be involved with difficult-to-describe phenomena. The transistor was used in circuits long before the internal behavior could be described in precise mathematical terms, and many parameters are still determined empirically. Consequently the proper perspective of fluidics' state-of-the-art requires a review of the histories and methods of these other technologies.

1.3 REVIEW OF THE ELECTRON TUBE TECHNOLOGY

The invention of the vacuum triode stimulated the development of a whole new area of the mathematical sciences. Because of its unique capabilities for control and amplification and its very high speed of response, it was employed in many new and sophisticated applications. There was so much pressure to make it do so many jobs that there were not enough highly trained and experienced electrical craftsmen who could design successfully by intuition to satisfy the demand. An analytical approach that would allow a great number of technicians to produce effective designs was an absolute necessity.

The technology, as it developed, definitely was functionally oriented. That is, the external behavior of the triode was defined mathematically, for example, $i_b = f(e_c + e_b)$, but its internal parameters (such as μ) were determined empirically, avoiding the need to develop a precise analytical description of the internal phenomena involved in the control of the electron beam.[2]

Since the vacuum triode is a nonlinear circuit device, its operating characteristics were found to be most conveniently expressed in the form of graphs or families of curves that indicate the actual experimental performance of the tube (volt-ampere relationships) as illustrated in Figure 1.2. In addition to using these graphical characteristics for circuit design — the technique will be described later — these curves assist in the determination of certain important performance parameters such as plate resistance, r_p, and amplification factor, μ, illustrated in Figure 1.2. These values are, in turn, used in equivalent circuit performance analysis.

The transfer curve of the vacuum triode is not linear over its whole course, but it can be considered to be linear over a small segment of the curve. Consequently, if a small signal is applied, the tube's operation is confined to only a small portion of the curve and thus may be considered linear. But for large signals, this assumption is invalid because a greater part of the curve is used. The result of this was that two distinct methods of analysis of vacuum tube circuits had to be developed; one that assumes linearity and is appropriate only for the consideration of small signals, and the other that includes the effects of nonlinearities and therefore is suited for large-signal conditions.

Graphical methods are ordinarily used for large-signal performance analysis to account for nonlinearities in the triode. They also are applicable to small-signal analysis, but because of the difficulties of including frequency-sensitive behavior and interpolating between curves, the inaccuracies and inconvenience are generally unacceptable. The "load line" method is the most often used type of graphical analysis (see Figure 1.3). The

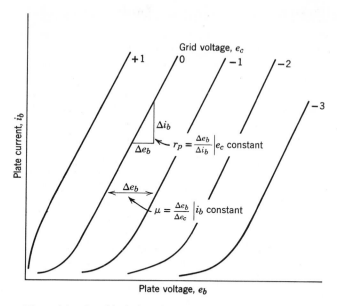

Figure 1.2 Graphical plate characteristics of a vacuum triode.

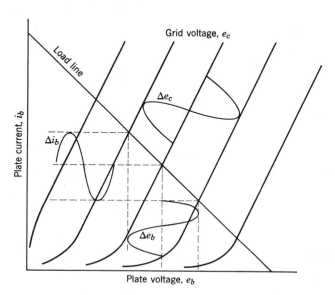

Figure 1.3 The load line method of circuit analysis.

5

curve representing the load characteristic is superimposed upon the plate characteristics of the triode. Since the tube and the load are connected in series and the individual curves represent unique operating conditions for each device, the only stable operating points are located where the curves intersect. Therefore, the plate voltage, e_b, for any value of grid voltage, e_c, can be determined from the points of intersection. Using this data, the voltage transfer curve can be plotted.

For small-signal and dynamic analysis, the equivalent circuit method is more convenient. Complex networks and high frequency behavior can be considered with relative ease using this approach. Figure 1.4 illustrates a real vacuum tube amplifier circuit and its linearized equivalent model. The fixed supply voltages E_{cc} and E_{bb} are eliminated because only incremental signals are considered. The tube can be replaced by a simple voltage generator, μ, and an equivalent internal resistance, r_p. As noted in Figure 1.2, the equivalent circuit parameters are determined from the graphical characteristics. Since the equivalent circuit concept is based on the assumption of linearity and fixed tube parameters, the values of μ and r_p can be determined at the point on the curves at which the circuit is to operate.

The general amplifier equivalent circuit must be altered for special operating conditions. For high frequency signal analysis, one must be concerned about the frequency-sensitive parameters of the vacuum tube and external circuit. The equivalent circuit for high audio frequencies is shown in Figure 1.5.

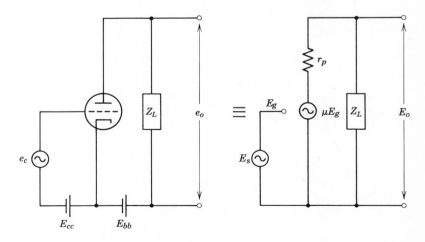

Figure 1.4 Derivation of the equivalent circuit for the vacuum tube amplifier.

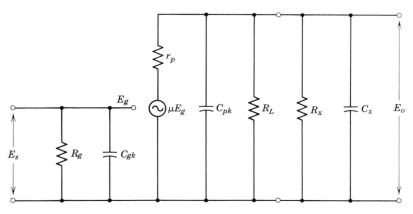

Figure 1.5 High-frequency equivalent circuit for the vacuum tube amplifier.

1.4 REVIEW OF THE PNEUMATICS TECHNOLOGY

One of the earliest applications of equivalent electrical circuits outside electronics was in the field of acoustics. The appropriate analogous elements are shown in Figure 1.6. Complex networks such as the automobile muffler (Figure 1.7) were analyzed and synthesized with good success by using these lumped parameter models.

More recently Dunn developed an electrical equivalent model of the human vocal tract (Figure 1.8). Dunn's model was the first step toward the representation of the distributed nature of the vocal tract as a series

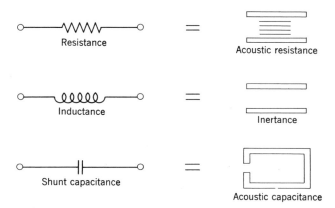

Figure 1.6 Analogous circuit elements.

8 History of Fluidic Systems Design Techniques

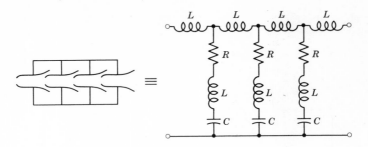

Figure 1.7 Automobile muffler and equivalent electric circuit.

of cylindrical sections, or acoustical lines, which makes it possible to use transmission line theory in finding its resonances. This initial model was able to produce acceptable vowel sounds and was used in research on the phonetic effects of articulator movements. As simple as it was, the model pointed the way for much subsequent research and has been greatly expanded.

Several considerations are appropriate to extend the equivalent circuit approach from acoustics to conventional pneumatic systems. One must be concerned with the effects of:

1. Constant velocity flow.
2. Heat dissipation.
3. Long lines.

The effective capacitance in pneumatic circuits has been found to be most nearly isothermal rather than adiabatic because the rates of compression are relatively slow and there is usually constant through-flow and considerable thermal mass in the system hardware; that is,

$$C = \frac{\text{volume}}{\text{absolute pressure}}$$

Long lines have presented a difficult analytical problem for many years. However, for many practical cases they can be represented by a distributed parameter model with as few as two or three sections.

As a result, successful methods of analysis for pneumatic systems which range from process controls to autopilots[17] have been developed in recent years. Increasingly, mechanical systems also have been studied with the impedance approach, using equivalent circuit analysis. Naturally, this allows the analysis of combinations of pneumatic and mechanical systems using a single coupled equivalent circuit.

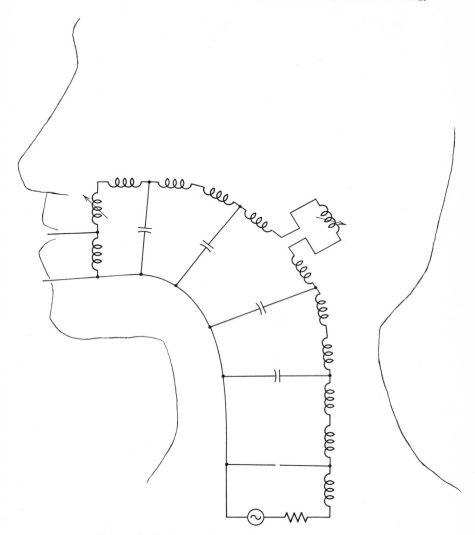

Figure 1.8 Equivalent electric circuit for human vocal tract.

1.5 REVIEW OF THE HYDRAULICS TECHNOLOGY

Because hydraulics are considered to be superior for fast response and high power, and because military aircraft and ships have used them almost exclusively, hydraulic systems have received a great amount of analytical attention since 1940.

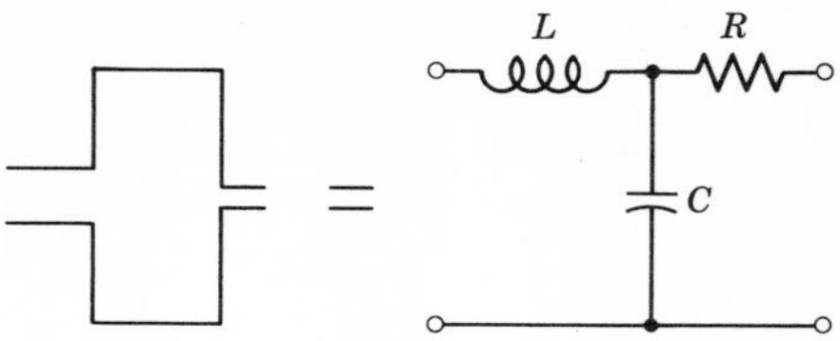

10 History of Fluidic Systems Design Techniques

Figure 1.9 Simplified lumped-parameter model for fluid conduit.

Some rather sophisticated methods have been used in these analyses. For example, in the field of fluid power transmission,[18] graphical characteristics are used to determine the effects of steady-state pipe losses on the system characteristics, and lumped parameter equivalent circuits with inertance, capacitance, and resistance are utilized in the analysis of dynamic characteristics as illustrated in Figure 1.9.

Whenever the significant wave lengths of the variables are small compared with the system's physical dimensions, the distributed nature of the system cannot be represented adequately by a lumped parameter network. In this case, distributed parameter representations composed of numerous sections are used, which are similar to those shown in Figure 1.9. The solution to the behavior of such a system is complex and often requires the aid of an electronic computer.

Most engineers compute the dynamic characteristics of relatively complicated hydraulic control systems by using various kinds of analogies such as hybrid hydraulic-mechanical equivalent circuits. Figure 1.10

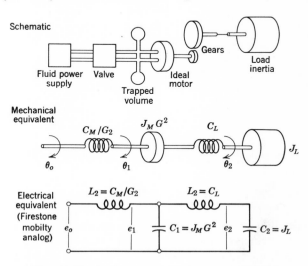

Figure 1.10 Analogous two-mesh systems (Ref. 18).

illustrates two different types of equivalent circuits for a valve-controlled hydraulic system. If systems are more complex and an electronic computer can be used in the analysis, then the engineer is increasingly apt to develop a complete mathematical description of the entire system and solve the resulting set of equations on the computer. This is, of course, the ultimate goal for systems analysts—to understand all variables and nonlinearities well enough so that they may be described mathematically.

1.6 REVIEW OF THE TRANSISTOR TECHNOLOGY

The third major event that stimulated the development of more sophisticated component and circuit analysis techniques was the introduction of the transistor. Again, circuit analysts found themselves under tremendous pressure to develop techniques that would allow the introduction of the device into many applications as quickly as possible. But they were relatively unprepared because of two very significant differences between vacuum tubes and transistors. The vacuum tube grid current is assumed to be negligible (infinite input impedance) and no feedback current exists from the plate to the grid circuit (zero reverse transfer admittance). The transistor, on the other hand, has significant input, output, and forward and reverse transfer admittances; therefore, it must be considered to be a coupled circuit in all directions. Consequently, the fifth stage of an amplifier must be considered in the analysis of the first stage. The potential complexity introduced by this requirement was staggering.

The accelerated development of techniques associated with the analysis of multiport network models alleviated the anticipated difficulties. It was found that in most circumstances the transistor could be represented adequately by a linear two-port model such as in Figure 1.11.[20] The two-port model is suitable because there is normally only one significant independent output variable.

When such a model can be considered to be linear, and when internal cross coupling is present such as in the transistor, one can write

$$E_1 = h_{11}I_1 + h_{12}E_2$$
$$I_2 = h_{21}I_1 + h_{22}E_2$$

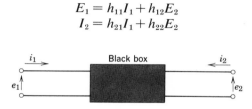

Figure 1.11 Two-port model and its variables.

where the h's are the driving-point and transfer hybrid parameters of the two-port model. If a unit voltage is applied to port 2, $h_{12}E_2$ is the complex voltage that appears at port 1. Using this set of hybrid parameters, the transistor equivalent circuit can be defined as shown in Figure 1.12.

However, no single model of a transistor is suitable for *all* circuit applications. As Linvill and Gibbons correctly state,[20] an attempt to account for all of the interacting factors with one model will needlessly complicate the analysis of even simple circuits. The circuit designer's needs are most adequately met by a *range* of models of varying complexity, including both simple and imperfect models that may be suitable representations of only the most dominant characteristics, and other more adequate but complicated models for situations where ultimate design limitations are to be studied.

Models can be classified according to two basic forms or concepts, physical or functional. Physical models relate terminal behavior to *internal* processes. For transistors, this type of model is developed from an analysis of the physical process that occurs within the semiconductor material. Functional models represent only abstract terminal properties and neglect the details of physical processes. These models are determined by *measuring* the response of the transistor to certain input signals.

The functional type of model is generally most appropriate since it provides sufficient information to predict the response of the device regardless of the particular excitation or the circuit in which it is connected. The empirical parameters for use with the model are monitored continually as a part of the quality control process in manufacturing. Their values can be measured directly with relatively simple instruments, or calculated by the designer from static characteristic curves.

Using the exact procedure employed with vacuum tubes, the operating point of the transistor can be determined by superimposing a load line on a plot of the static output characteristic curves. However, transistor characteristic curves vary widely with temperature and among individual

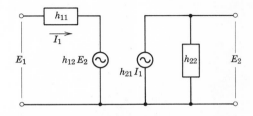

Figure 1.12 Equivalent two-port representation of a transistor.

units of a given type; therefore, somewhat different forms of the static characteristics must be used to ensure the proper bias for maximum stability. A linear circuit solution that involves linear approximations of the static characteristics is often satisfactory for the establishment of the bias point of the transistor.

1.7 REVIEW OF NONLINEAR CIRCUIT ANALYSIS

A comprehensive description of the state-of-the-art of nonlinear circuit analysis is far beyond the scope of this book. However, a brief review of the technology is useful to indicate several important factors in the analysis of fluidic devices.

The electronic computer has made a very significant impact upon nonlinear systems analysis. The behavior of nonlinear components and simple circuits have long been analyzed using piecewise linear techniques. But the applicability of this approach has been limited because the conditions for transition from one linear piece to the next have been difficult to determine manually. Now the introduction of the electronic computer makes the transition solution easier, and piecewise linear analysis has become a powerful tool for the designer.

Analog computers provide another approach to nonlinear systems analysis. Using nonlinear function generators, one can simulate most nonlinear and multivalued functions that govern the behavior of complex systems. Consequently, total system simulation can be accomplished.

Methods for the analysis of highly complex coupled networks and structures have been developed in recent years. It is now possible to describe extensive systems with nonlinear parameters (systems such as thin shells and inflatable structures) by using matrices of high order to treat a distributed system as a very large network of lumped elements. Again, until the introduction of the electronic digital computer, it was impractical to attempt such solutions. However, we now have the tools necessary to describe complex multiport networks with matrices and solve systems of interconnected matrices using the electronic digital computer.

1.8 SUMMARY OF THE REVIEW OF OTHER TECHNOLOGIES

Up to this point, we have explored various other technologies to show that they employ very similar techniques in component description and systems analysis. These general techniques can be reduced to three basic forms:

1. Mathematical (linear and nonlinear).
2. Graphical.
3. Linear models.

Recently the emphasis has been upon developing more sophisticated linear models by introducing greater precision in treating additional degrees of freedom and more comprehensive mathematical descriptions that incorporate nonlinearities and a large number of degrees of freedom.

The preceding history of the development of other technologies will be used as a basis for measurement of current progress in fluidic systems and predictions about the future directions of accomplishment.

1.9 REVIEW OF THE FLUIDICS TECHNOLOGY

Hopefully the previous discussion has placed fluidics in its proper perspective. Fluidics should not be considered a radically different field such as the development of mathematics as compared to psychology, rather it is a new technology comparable to the introduction of semiconductors after vacuum tubes. Fluidics is simply another system obeying firm physical laws, and not the semimystical "black art" that some would have us believe. Consequently, fluidic systems can be analyzed, described, and synthesized with many of the tools used for more familiar physical systems.

Fluidic systems analysis is not simple. However, the basic tools that are required are available—no new "breakthroughs" are necessary. It is simply a matter of time until these tools can be adapted for most effective usage.

Prior to its introduction, work that was significant to fluidics was accomplished in the related field of pneumatics. In 1950, A. S. Iberall published a paper[5] about the transmission of oscillatory pressures in instrument lines. The importance of the paper to fluidics was its concentration on oscillatory pressures of relatively low frequencies (well below acoustic frequencies) and levels of oscillation that are an appreciable fraction of the mean pressure. These conditions are directly relevant to fluidic circuits except that through-flow is absent. During the same period in which Iberall's paper was published, analog and digital computers that provided much increased analytical capabilities, especially for complex systems, were becoming commonplace. Consequently, considerable discussion and interest in the numerous analogies between electrical, mechanical, and hydraulic systems was evident.[6–9]

One of the first applications of the electric circuit analogy to pneumatic systems was presented in "On the Dynamics of Pneumatic Transmission Lines" by Rohman and Grogan, which was published in May 1957.[11] Their analysis was linearized to create a form that is convenient for engineering computation of complete systems dynamics. Pressure was established to be analogous to voltage and volumetric flow to current. The pneumatic equivalents of resistance, capacitance and inductance were also introduced. The authors showed experimental data to prove that their equivalent circuit, shown in Figure 1.13 was a valid representation of the pneumatic circuit. One of their very important conclusions was that the isothermal rather than the adiabatic assumption for capacitance is much more realistic.

In November 1957, Ezekiel and Paynter[12] introduced the systems concept to the analysis of fluid systems. They showed that individual transducers, modulators (valves), transmitters, and passive elements with resistance, capitance, and inertance could be represented as two-port networks. Thus, whole systems could be fabricated in this form, and thus could easily be programmed for solution on an electronic computer.

The application of electric equivalent circuits to the analysis of pneumatic systems was firmly entrenched by 1958. Various textbooks which were published at that time described equivalent circuits for flapper-nozzle valves and numerous filter networks (a typical example is shown in Figure 1.14).[13,14]

Modern concepts that are employed in fluidic devices were first invented in 1959. Following the announcement by Diamond Ordnance Fuze Laboratories of the work of Bowles, Warren, and Horton, numerous companies and government agencies began their attempts to apply fluidic devices in useful systems. A few of these groups and individuals were partially successful, but the great majority (including the author) were

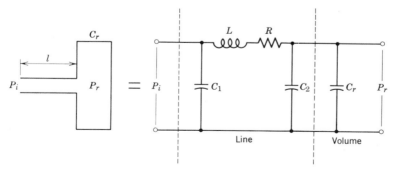

Figure 1.13 Electric equivalent of a pneumatic circuit (Ref. 11).

16 History of Fluidic Systems Design Techniques

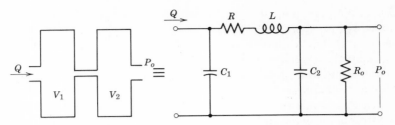

Figure 1.14 Equivalent circuit for pneumatic low-pass filter network.

frustrated by the lack of uniformity in the description of performance of fluidic devices and the need for suitable methods to couple these devices into efficient circuits. It soon became apparent that the lack of adequate analytical methods for both internal and external description of the fluid amplifier would seriously handicap the advancement of the technology.

A serious problem was the difficulty that was encountered in cascading amplifiers. Warren described some of the considerations involved in cascading fluidic devices and proposed a solution in a paper published in August 1962.[24] He introduced the use of a bleed port as the most direct method for matching cascaded stages, by avoiding the need for accommodating all the flow of preceeding stages.

Several of the papers presented at the First Fluid Amplification Symposium held at the Diamond Ordnance Fuze Laboratories in October 1962 indicated the need for systems analysis methods and described various aspects of the associated problems. Authors only skirted the fringes of systems synthesis; the nearest approach was a recognition of impedance matching requirements and the potential usefulness of fluid to electrical analogies. Katz described proportional amplifiers and suggested the use of the electric analogy for circuit elements.[28] Hicks and Jetter stated the importance of demonstrating analogies between pneumatic and electric systems so the established electrical circuit theory could be used to predict behavior.[25] The authors tabulated analogies between units and parameters of mechanical, electric, and pneumatic systems, and demonstrated some analytical methods for circuits that contain passive elements. Lechner and Wambsganns described the output and input impedance characteristics of fluidic amplifiers and their significance to the cascading problem.[26] Dexter described an "admittance" parameter, Q^2/P, for use in classifying and matching components.[27]

The American Society of Mechanical Engineers held a Symposium on Fluid Jet Control Devices in November 1962 and considerable activity in systems synthesis was reported. Norwood's paper was most significant since it was the first to view the fluidic amplifier as a multiport model or "black box".[34] He also examined the switching behavior of the bistable amplifier with a series restriction in the control line by defining the input impedance curve and superimposing it upon the orifice output pressure-flow characteristic. Brown reported work with Van Koevering in which load-flow characteristics of fluid amplifiers were measured experimentally and then applied graphically to the problem of matching the fluid amplifier with a load.[30] Dexter extended his "admittance" concept and showed the graphical transfer and output characteristics illustrated in Figure 1.15.[31] Warren also utilized graphical output characteristics, P_o versus Q_o, to describe the performance of various bistable amplifiers.[35] Boothe concluded in his paper that "the characteristics of fluid amplifiers can fall into two general categories, namely input and output characteristics. From these characteristics the transfer characteristics or gains can be deduced."[33]

In August 1963, Katz and Dockery published the first comprehensive work on static component description and circuit synthesis.[37] This report was a major milestone in fluidics technology. Both proportional and digital devices were considered, using techniques for component description and also system design. Static characteristic curves were employed throughout the study and tremendous insight into the behavior of fluidic devices was provided by illustrations of the variation of the characteristics that were presented.

The proportional fluidic amplifier was represented as a four-terminal network. Pressure gain and the pressure amplification factor, that is, the pressure gain at constant output flow, were defined. A mathematical expression for the static output characteristic was developed and used to illustrate the effect of varying parameters. An expression also was developed for the input characteristics to illustrate the effect of variation.

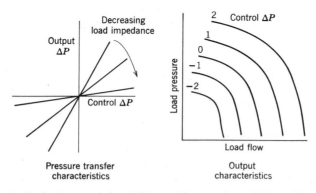

Figure 1.15 Static characteristics of differential proportional fluid amplifier (Ref. 31).

These input and output characteristics were applied to the analysis of the behavior of two cascaded amplifiers. A matching procedure was outlined that satisfied the criterion of simultaneous saturation.

The bistable amplifier was treated similarly. Output characteristics and the effect of parameter changes were illustrated. Again the load line technique was applied to the analysis of cascaded stages.

Katz and Dockery took a commendable position in the conclusion to their study which, unfortunately, was not appreciated by the majority of people in fluidics for several years. They stated that "the fluid-flow phenomena in fluid amplifiers are extremely complex. *However, to use the amplification phenomena it is not necessary to understand the flow.* It is necessary, though, to have a method of measuring performance so that amplifier circuits can be designed. The black-box approach and the resulting characteristic curves presented here provide one such method."[37] It is unfortunate that even today, 7 years later, manufacturers are reluctant to provide these characteristic curves for their fluidic devices.

Further work on component description and systems analysis was not published in the open literature until May 1964, when numerous related papers were presented at the Second Fluid Amplification Symposium at Harry Diamond Laboratories. Many significant results were reported by the participants.

Belsterling and Tsui reported progress on a research program devoted exclusively to developing techniques for fluidic system analysis and synthesis.[46] They extended the graphical approach for static analysis and originally applied the equivalent circuit approach for dynamic analysis of proportional fluidic amplifiers. Consequently, the authors took the initial step in the development of dynamic system design methods which are still the only practical techniques. The equivalent circuit for the vented jet-interaction amplifier that is operating at practical (intermediate) frequencies is shown in Figure 1.16. The authors reported successful experimental verification of the equivalent circuit for both the small-signal static case and for one dynamic case.

Brown presented an analysis of fluid systems as coupled multiport elements.[44] The amplifier was treated as a four-port element consisting of two control ports and two output ports. Supply and exhaust ports were neglected since no active signal is sent through them. Two methods for predicting stability were outlined, one employed admittance matrices, and the other used scattering variables. Some examples that illustrated the application of these methods and suggested experimental techniques for measuring the admittances and scattering operators were given.

Brown's paper was the beginning of another new approach to fluidic systems analysis. The application of the general systems approach to this

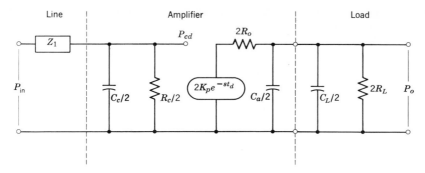

Figure 1.16 Midfrequency equivalent circuit for vented jet-interaction proportional fluid amplifier (Ref. 46).

new technology implied a recognition that fluidics, because it is similar to numerous other phenomena, does not, indeed, require extraordinary techniques.

Similar techniques were applied to new fluidic devices. Lechner and Sorenson used graphical techniques to describe and assist the cascading of impact modulators.[49] The input impedance curve was used as a load line on the graphical output characteristics to determine the operating points. However, because of the presence of the second independent supply nozzle, the analytical process for impact modulators involves a second set of output characteristics. Mayer and Maker developed a mathematical expression for the performance of the vortex valve and illustrated output characteristics with a superimposed load line.[41]

Extensions of the general techniques were also reported. Saghati illustrated the application of the "black box" graphical and equivalent circuit concept to the design of fluidic logic circuits.[42] Katz, Goto, and Dockery reported considerable success in the design of analog computing circuits with a straightforward analytical approach that was based on the use of performance parameters in addition to analogous electrical circuits.[50] Warren described graphical methods for interconnecting bistable fluidic amplifiers by superimposing input characteristics on the output characteristics.[45]

Two papers that are significant for fluidic systems analysis were presented at the Joint Automatic Control Conference in June 1964. Wright[53] extended Norwood's work[34] with graphical design techniques to ensure proper matching. A stability criterion was also discussed. Boothe presented a method for predicting the dynamic response of a proportional fluid amplifier using a lumped parameter representation.[52] To validate the approach, frequency response tests were run with air and water up to several kilohertz. Experimental data for the tests using air correlated well with the analytical results, even at high frequencies where the lumped-parameter approach might be expected to break down. Boothe's paper represents another major milestone in fluidic systems analysis because the fluidic circuit was analyzed in a conventional manner and experimental data proved that the lumped-parameter representation was valid for most practical cases using any operating fluid.

However, reviews of fluidic techniques disagreed about the eventual importance of the lumped-parameter approach. Shinn and Boothe[54] described static matching of cascaded fluidic devices using characteristic curves and recommended the application of the lumped-parameter circuit model for situations where the operating frequencies are below a few hundred hertz. Fox and Wood,[55] in a summary of the state of the technology published in October 1964, pointed out that the fluid circuit is extremely complex and cannot generally be treated as a lumped-parameter system. They recognized that simplifications such as lumped-parameter equivalents and graphical characteristics must be used for the present, but stressed that simultaneous solution of the fluid flow equations for the entire system must be the ultimate goal.

Also in October 1964 the staff of General Electric company completed their state-of-the-art studies for NASA. The need for standardization of parameters to describe the behavior of fluidic amplifiers and uniform methods for parameter measurement was emphasized. The studies indicated that time delays in proportional amplifiers resulted from three sources:

1. Compressibility and inertance in the control ports.
2. Transport time from the nozzle to the receiver.
3. Receiver and load dynamics.

Suggested staging techniques were presented that described the fluid amplifier as an *n*-port network defined by an admittance matrix. Stability analysis — for small disturbances, this is an extension of the admittance matrix methods — with nonlinearities was illustrated using graphical methods. The latest useful staging techniques were described. These methods were basically graphical and ignored cross-coupling effects (reverse transfer admittances).

Fluidics, a comprehensive study of the technology, contains several discussions important to the subject of fluidic systems.[58] Graphical load line methods for analyzing both proportional and bistable amplifier circuits are shown. Dynamic analysis with linear equivalent circuits is also discussed. A technical appendix to this book contains several significant papers.

In the appendix, Fox reviews mathematical approaches to the analysis of the behavior of fluidic devices. He reported that with the aid of electronic computers, some progress was being made in solving simple two-dimensional flow problems. Fox described the analogies between fluid flow and electric circuits. He states that "there is an exact analogy between elementary fluid theory and simple lumped-parameter electrical circuit theory. Other analogies such as the acoustic analogy with the telegraph equations (distributed parameters) also exist and have usefulness within the limits of the analogy." Further *"useful results can be obtained by proper use of these analogies and these results are applicable to design problems"* (italics mine).

Also in the technical appendix of *Fluidics*, Brown evaluated the future of the analysis of fluidic systems. He defined three types of models:

1. Mathematical (equations).
2. Functional (diagrams).
3. Reticulation (equivalent models).

Bondgraphs and pure-delay models were described. In a review of current knowledge, Brown concentrated on the equivalent circuit methods, which were just beginning to show promise. In conclusion, he stated "extremely useful prime-element dynamic models of linear and nonlinear fluid amplifiers and logic elements should be achievable..." and "lumped, pure delay and mixed models offer different compromises between the physical feeling they impart and the ease with which they can be realized computationally. All, however, are superior to mathematical or signal-flow-diagram representations." He predicted that the first use of the models would occur in systems synthesis, and the second use in focusing attention on the phenomena that limit the dynamic performance of a component or system.

Brown's comments constitute another important milestone. He established the perspective for viewing the work then being done against the general field of dynamic analysis. Consequently, he provided the direction for future work.

Goldschmeid et al.[60] published in July 1965 a comprehensive report on the analysis of fluid amplifier dynamic characteristics. They had developed the analogies between fluid and electric circuits and used them to illustrate several equivalent circuits. But they cautioned the reader on the use of lumped parameters throughout a system. Fox and Goldschmeid, in an appendix to the report, show their analysis which led to the conclusion that electric analogies are only suitable for acoustic (a-c) signals and are useless for dealing with steady flow (d-c), which is a recognized limitation on linearized equivalent circuits.

Additional information on the methods for analyzing proportional fluid amplifier circuits was published in August 1965 by Belsterling and Tsui.[61] Graphical methods for static and large signal analysis were illustrated in detail. Equivalent circuits for the proportional jet-interaction amplifier were shown for low frequency and high frequency (400 Hz) signals. Good correlation between predicted and experimental frequency response was obtained. A prediction of the effect on frequency response of a change in supply pressure also was found to be in good agreement with the actual data.

In September 1965, Kirshner published a brief resume of fluid circuit theory.[62] The fluid-electric analogy was explored in detail and the conclusion was drawn that the current analog should be mass flow and the voltage analog should be $\int(dp/\rho)+(v^2/2)$. Since this mechanical potential, as he calls it, is dependent on the stage path, nonlinearities due to conversion of energy into heat are present. This result leads to the conclusion that the fluidics engineer will always require graphical and computer techniques to consider these nonlinearities in his analysis of circuits and systems.

The Third Fluid Amplification Symposium produced, as did past symposia, a number of relevant papers. Especially important for the present review was the study by Roffman and Katz.[63] Their work concerned an experimental and analytical investigation of the application of conventional control system stability analysis to a fluidic circuit. They measured open-loop frequency response, predicted the closed-loop stability using the Nyquist criterion, then closed the loop experimentally and compared results. The correlation of the analytical to experimental results was good.

Belsterling published the results of a large-scale analytical and experimental study on static and dynamic analysis in February 1966.[67] The report covered the continuing development of analytical techniques for the design of systems of fluidic components. Three widely different amplifiers that were analyzed according to previously developed methods are as follows:

1. Vented jet-interaction type.
2. Closed (nonvented) jet-interaction type.
3. Vented elbow type.

Electrical equivalent circuits were derived and the frequency response predicted for each type. The analytical results correlated well with experimental tests. The major significance of this work was that various types of complex amplifiers were successfully represented by equivalent circuit models. The equivalent circuit for the closed type of amplifier (Figure 1.17) included the phenomena of internal feedback (reverse transfer admittance). This was the first successful attempt to analyze such a condition.

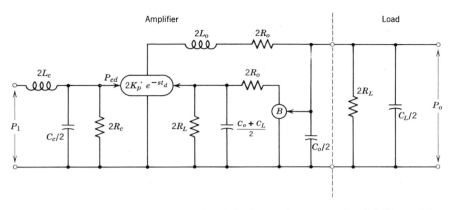

Figure 1.17 Equivalent circuit for closed jet-interaction proportional fluid amplifier (Ref. 67).

Another book, *Fluid Amplifiers*, edited by Kirshner and written by the staff of Harry Diamond Laboratories contains information relevant to this evaluation of fluidic systems analysis.[66] A chapter on fluid circuit theory contains an excellent discussion of the analysis of two-terminal pairs (two-port networks), fluid transmission lines, and the matching of lines with bleeds, vents, and branches. Beam deflection amplifiers were defined as two-terminal pairs and performance parameters were derived for analyzing dynamic performance. Characteristic curves for both proportional and bistable amplifiers were illustrated together with procedures for matching, using the load line technique.

In April 1966, Belsterling published an extension of the work on the development of systems design methods.[68] The article described the application of graphical techniques and equivalent electrical circuits to the analysis and synthesis of systems. The particular system that was used as an illustrative example is shown in Figure 1.18. Static behavior of the coupled components was determined by graphical characteristics and load lines. Dynamic performance was obtained by using the electrical equivalent circuit for the complete system. This presentation was the first detailed description to be published of the entire procedure for straightforward systems design.

A new equivalent circuit model for fluidic transmission lines was reported in December 1966 by Karam.[69] He showed by correlation with experiment that at low frequencies the line parameters can be considered to be independent of frequency, but at the high frequencies that are encountered in many fluidic circuits, the variation in the line parameters with frequency (including the resistance parameter) must be taken into account.

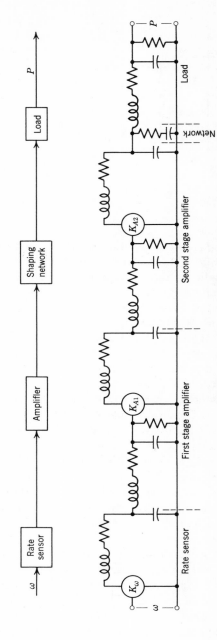

Figure 1.18 Equivalent circuit for complete illustrative system (Ref. 68).

24

At the May 1967 ASME Fluidics Symposium, Manion[70] presented a paper reporting on analog computer simulation of the proportional fluidic amplifier that illustrates one more very powerful technique for analyzing and synthesizing fluidic systems that has not yet been used to its greatest potential. With the aid of the analog computer, linear and nonlinear behavior both statically and dynamically can be simulated. The analog computer approach to fluidic systems design represents an important step with great promise for the future. The simulation model is shown in Figure 1.19, where Z_r and Z_v have the characteristics of the receiver and vent, respectively.

In July 1967, Belsterling published a comprehensive "how-to-do-it" guide to the design of analog systems using fluidic devices.[71] Reference information was presented for definitions, symbols, and general principles, an integrated set of static and dynamic design methods, and a step-by-step design procedure, illustrated with a practical sample problem—the design of a fluidic yaw damper system for the UH-1B helicopter. Component descriptions and systems synthesis methods were based on graphical characteristics and electrical equivalent circuits.

This work presented analog fluidic systems design procedures in minute detail, including methods applicable for sensors, amplifiers of various

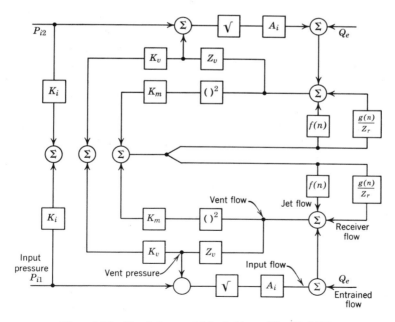

Figure 1.19 Simulation model for fluid amplifier (Ref. 70).

types, and actuators. Using the techniques described, the control systems engineer could treat fluidic elements in the same way as electronic, electromechanical, and hydraulic devices in the design of integrated systems. Insight into the functional behavior of fiuidic devices was also provided without dependence on the principles of fluid mechanics. It provided the systems designer with the analytical tools to produce working fluidic systems immediately. The report also provides the knowledge necessary to produce fluidic devices which are capable of good systems performance.

In July 1967 Pill reported a random testing technique which is especially suited to measuring dynamic transfer functions (or admittance matrices) at small-signal levels and at high frequencies.[72]

A major step in the direction of more accurate and more efficient fluidic systems analysis, indeed, the direction of the future, was the publication of an analytical and experimental study by Brown of the multiport network approach to fluidic systems synthesis.[73] The linearized characteristics of fluidic amplifiers including the self-admittances of each port and the transfer admittances between ports were represented by admittance matrices. Stability criteria and response relations were given in terms of these matrices. The frequency-dependent elements of the matrices were measured for a proportional jet-interaction amplifier using two experimental techniques.

Based on the reviews of other technologies and the progress in fluidics to date it is clear that multiport representations of fluidic devices must be used if the complex (and real) coupling between individual ports is to be included in the analysis. Brown's work provides the first real "handle" to doing this, covering the matrix description of the proportional fluidic amplifier, and the application of the matrix technique to stability criteria, and computation of response. Of particular significance in the study was the somewhat detailed description of the practical problems associated with methods for the direct measurement of the terms of the admittance matrices.

In October 1969 Brown and Humphrey[77] published the results of tests on a large-scale proportional amplifier. These tests defined the eight admittance terms of a four-port model over the entire useable frequency spectrum. The four admittances of a two-port (push-pull) representation were also given, which showed that the reverse transfer admittance term, as expected in a properly vented amplifier, can be neglected.

A rather brief general review of the history of fluidic systems synthesis techniques has been presented. The status of the development of such techniques has been established for the present purposes.

BIBLIOGRAPHY*

1. H. F. Olson, *Dynamic Analogies*, D. Van Nostrand, New York, Inc., 1943.
2. R. Ryder, *Electronic Engineering Principles*, Prentice-Hall Inc., New York, 1947.
3. E. A. Guilleman, *Mathematics of Circuit Analysis*, John Wiley & Sons, Inc., New York, 1949.
4. G. Murphy, *Similitude in Engineering*, Ronald Press, New York, 1950.
5. A. S. Iberall, "Attenuation of Oscillatory Pressures in Instrument Lines," *Research Paper RP2115*, National Bureau of Standards, 1950.
6. S. Goldstein, *Modern Developments in Fluid Mechanics*, Vol. I, Claredon Press, Oxford, 1950.
7. J. C. Schoenfeld, "Analogy of Hydraulic, Mechanical, Acoustic and Electric Systems," *Appl. Sci. Res., Sect. B*, **3** (1951–1953), pp. 417–450.
8. H. M. Paynter, "Electrical Analogies and Electronic Computers—Surge and Water Hammer Problems," *Trans. Amer. Soc. Civil Eng.*, **118** (1953), pp. 962–1009.
9. W. W. Soroka, *Analog Methods in Computation and Simulation*, McGraw-Hill Book Co., New York, 1954.
10. R. P. Bigliano, "Here's a Way to Measure Pneumatic Component Dynamics," *Control Eng.*, August 1956.
11. C. P. Rohmann and E. C. Grogan, "On the Dynamics of Pneumatic Transmission Lines," *Trans. ASME*, **79**, no. 4 (May 1957).
12. F. D. Ezekiel and H. M. Paynter, "Computer Representations of Engineering Systems Involving Fluid Transients," *Trans. ASME*, **79**, (November 1957), pp. 1840–1850.
13. J. E. Gibson and F. B. Tutuer, *Control System Components*, McGraw-Hill Book Co., New York, 1958.
14. D. P. Campbell, *Process Dynamics*, John Wiley & Sons, New York, 1958.
15. M. P. Risetnbatt and R. L. Riddle, *Transistor Physics and Circuits*, Prentice-Hall, Inc., Englewood Cliffs, N. J., 1958.
16. R. E. Norwood, "Pneumatic Flapper Valve Study," *Proc. First International Federation of Automatic Control*, Moscow, June 1960.
17. C. A. Belsterling and A. Marmarou, "Analysis of the Dynamics of the Tactair T3 Autopilot," *Franklin Institute Report I-A2212-02-1*, prepared for Tactair Division, Aircraft Products Co., Bridgeport, Pa., July 1960.
18. J. C. Blackburn, G. Reethof, and J. L. Shearer, *Fluid Power Control*, John Wiley & Sons, Inc., New York, 1960.
19. Robert E. Raymond, "Electrohydraulic Analogies—Systems Approach to Fluid Power Design," *Hydraulics and Pneumatics*, March 1961.
20. J. G. Linvill and J. E. Gibbons, *Transistors and Active Circuits*, McGraw-Hill Book Co., New York, 1961.
21. M. V. Joyce and K. K. Clarke, *Transistor Circuit Analysis*, Addison-Wesley Publishing Co., Reading, Mass., 1961.

*In chronological order.

22. C. L. Strong, "How Streams of Water Can Be Used to Create Analogies of Electronic Tubes and Circuits," *Sci. Amer.*, August 1962.
23. H. Diamond and S. Fenster, "Hydraulic Control System Analysis," *Electrotechnol.*, August 1962.
24. R. W. Warren, "Fluid Amplification 3—Fluid Flip-Flops and a Counter," *Report TR-1061*, Diamond Ordnance Fuze Laboratories, Washington, D.C., August 1962.
25. B. A. Hicks and E. S. Jetter, "Pneumatic Linear Circuits," *Proc. Fluid Amplification Symp.*, Diamond Ordnance Fuze Laboratories, Washington, D.C., October 1962.
26. T. J. Lechner and M. W. Wambsganns, "Proportional Power Stages for Impedance Matching," *Proc. Fluid Amplification Symp.*, October 1962.
27. E. M. Dexter, "A Technique for Matching Pure Fluid Components Applied to the Design of a Shift Register," *Proc. Fluid Amplification Symp.*, October 1962.
28. S. Katz, "An Introduction to Proportional Fluid Control," *Proc. Fluid Amplification Symp.*, October 1962.
29. D. P. Costa, "Electrical Analogs for Analysis of Electro-Acoustical Transducers," *Electromech. Design*, November 1962.
30. F. T. Brown, "A Combined Analytical and Experimental Approach to the Development of Fluid Jet Amplifiers," *ASME Symp. Fluid Jet Devices*, New York, November 1962.
31. E. M. Dexter, "An Analog Pure Fluid Amplifier," *Proc. ASME Symp. Fluid Jet Devices*, November 1962.
32. R. W. Warren, "Some Parameters Affecting the Design of Bistable Fluid Amplifiers," *ASME Symp. Fluid Jet Devices*, November 1962.
33. W. A. Boothe, "Performance Evaluation of a High Pressure Recovery Bistable Fluid Amplifier," *ASME Symp. Fluid Jet Devices*, November 1962.
34. R. D. Norwood, "A Performance Criteria for Fluid Jet Amplifiers," *ASME Symp. Fluid Jet Devices*, November 1962.
35. R. W. Warren, "Some Parameters Affecting the Design of Bistable Fluid Amplifiers," *ASME Symp. Fluid Jet Devices*, November 1962.
36. J. E. Fleckenstein, "Method for Determining the LaPlace Transform of a Pneumatic Nozzle-Flapper Combination," *ASME Paper 62-WA-15*, 1962.
37. S. Katz and R. J. Dockery, "Fluid Amplification II—Staging of Proportional Bistable Fluid Amplifiers," *Report TR-1165*, Harry Diamond Laboratories, Washington, D.C., August 1963.
38. C. A. Belsterling and K. C. Tsui, "Research on the Performance of Pure Fluid Amplifiers: Part I—Static or Low Frequency Case," *Franklin Institute Laboratories Interim Report I-09761-1*, August 1963.
39. F. T. Brown, "A Combined Analytical and Experimental Approach to the Development of Fluid Jet Amplifiers," *ASME Paper 62-WA-154*, December 1963.
40. G. L. Roffman, "Staging of Closed Proportional Fluid Amplifiers," *Proc. Second Fluid Amplification Symp.*, Harry Diamond Laboratories, Washington, D.C., May 1964.
41. E. A. Mayer and P. Maker, "Control Characteristics of Vortex Valves," *Proc. Second Fluid Amplification Symp.*, **2** (May 1964), pp. 61–84.
42. H. T. Saghati, "Static Design of Pneumatic Logic Circuits," *Proc. Second Fluid Amplification Symp.*, **2** (May 1964), pp. 191–220.

43. G. D. Roffman, "Staging of Closed Proportional Fluid Amplifiers," *Proc. Second Fluid Amplification Symp.*, **3** (May 1964), pp. 73–84.
44. F. T. Brown, "On the Stability of Fluid Systems," *Proc. Second Fluid Amplification Symp.*, **1** (May 1964), pp. 233–256.
45. R. W. Warren, "Interconnection of Fluid Amplification Elements," *Proc. Second Fluid Amplification Symp.*, **3** (May 1964), pp. 53–72.
46. C. A. Belsterling and K. C. Tsui, "Application Techniques for Proportional Pure Fluid Amplifiers," *Proc. Second Fluid Amplification Symp.*, **2** (May 1964), pp. 163–190.
47. W. L. Cochran and R. W. Tilburg, "The Staging of Pressure Proportional Amplifiers to Provide Stable, Medium Gain, Dual Control, Single Output Pure Fluid Systems," *Proc. Second Fluid Amplification Symp.*, **2** (May 1964), pp. 289–312.
48. R. W. Van Tilburg and W. L. Cochran, "Development of a Proportional Amplifier for Multi-Stage Operation," *Proc. Second Fluid Amplification Symp.*, **2** (May 1964), pp. 313–334.
49. T. J. Lechner and P. H. Sorenson, "Some Properties and Applications of Direct and Transverse Impact Modulators," *Proc. Second Fluid Amplification Symp.*, **2** (May 1964), pp. 33–60.
50. S. Katz, J. L. Goto, and R. J. Dockery, "Experiments in Analog Computation with Fluids," *Proc. Second Fluid Amplification Symp.*, **2** (May 1964), pp. 335–374.
51. R. L. Humphrey and R. M. Manion, "Low-Pass Filters for Pneumatic Amplifiers," *Proc. Second Fluid Amplification Symp.*, **1** (May 1964), pp. 252–278.
52. W. A. Boothe, "A Lumped Parameter Technique for Predicting Analog Fluid Amplifier Dynamics," *Proc. Joint Automatic Control Conf.*, June 1964.
53. C. P. Wright, "Some Design Techniques for Fluid Jet Amplifiers," *Joint Automatic Control Conf.*, June 1964.
54. J. N. Shinn and W. A. Boothe, "Connecting Elements into Circuits and Systems," *Control Eng.*, September 1964.
55. H. L. Fox and O. L. Wood, "The Development of Basic Devices and the Need for Theory," *Control Eng.* October 1964.
56. P. L. Abbey, et al., "Fluid Amplifier State of the Art," *NASA Report CR-101*, October 1964.
57. F. T. Brown, S. D. Graber, and R. E. Wallhagen, "Investigation of Stability Predictions of Fluid-Jet Amplifier Systems," *NASA Report CR-54244*, October 30, 1964.
58. E. F. Humphrey and D. H. Tarumoto (Eds.), *Fluidics*, Fluid Amplifier Associates, Boston, Mass., 1965, revised edition 1968.
59. F. T. Brown, "On the Future of Dynamic Analysis of Fluid Systems," in *Fluidics*, Fluid Amplifier Associates, Boston, Mass, 1965.
60. R. F. Goldschmied, et al., "Analytical Investigation of Fluid Amplifier Dynamic Characteristics," *NASA Reports CR-244 and CR-245*, **I** and **II**, July 1965.
61. C. A. Belsterling and K. C. Tsui, "Analyzing Proportional Fluid Amplifier Circuits," *Control Eng.*, August 1965.
62. J. M. Kirshner, "Some Topics in Fluid Circuit Theory," *Proc. First International Conf. Fluid Logic and Amplification*, Cranfield, England, September 1965.
63. G. L. Roffman and S. Katz, "Predicting Closed Loop Stability of Fluid Amplifiers from Frequency Response Measurement," *Proc. Third Fluid Amplification Symp.*, Harry Diamond Laboratories, Washington, D.C., October 1965.

64. J. M. Kirshner, "A Definition of the Mechanical Potential Necessary to a Fluid Circuit Theory," *Proc. Third Fluid Amplification Symp.*, October 1965.
65. E. C. Hind and E. J. Hahn, "The Transfer Function of the Pneumatic Capacitance," *ASME Paper 65-WA-AUT-18*, November 1965.
66. J. M. Kirshner (Ed.), *Fluid Amplifiers*, McGraw-Hill Book Co., New York, 1966.
67. C. A. Belsterling, "Development of the Techniques for the Static and Dynamic Analysis of Fluid State Components and Systems," *U.S. Army AVLABS Technical Report 66-16*, February 1966.
68. C. A. Belsterling, "Designing Fluidic Systems," *Control Eng.*, April 1966.
69. J. T. Karam, "A New Model For Fluidics Transmission Lines," *Control Eng.*, December 1966.
70. F. M. Manion, "Proportional Amplifier Simulation," *ASME Advances in Fluidics-1967 Fluidics Symp.*, New York, May 1967.
71. C. A. Belsterling, "Fluidic Systems Design Manual," *U.S. Army AVLABS Report 67-32*, July 1967.
72. Juri Pill, "Dynamic Characterization of a Proportional Fluidic Amplifier by the Use of Stochastic Signals," *Report EDC 7-67-19*, Case Western Reserve University, Cleveland, Ohio, July 1967.
73. F. T. Brown, "Stability and Response of Fluid Amplifiers and Fluidic Systems," *NASA Report CR72191*, October 1967.
74. W. P. Depperman, "Miniature Integrated Fluidic Circuits," *Fluidics Quart.*, **1**, no. 1 (October 1967).
75. W. T. Rauch, "Putting Fluidics to Work", *Fluidics Quart.*, **1**, 1 (October 1967).
76. C. A. Belsterling, "Digital and Proportional Jet-Interaction Devices and Circuits," *Fluidics Quart.*, **1**, no. 2 (April 1968).
77. R. A. Humphrey and F. T. Brown, "Dynamics of a Proportional Fluidic Amplifier; Part 1," *ASME Paper 69-WA/Flcs-2*, October 1969.

2
The Fluidic Systems Design Process

The purpose of this chapter is to introduce the systems design process as applied to fluidics, and to alert the reader as to what methods can be used to describe, test, and select components, and to design systems.

2.1 THE GENERAL APPROACH

The fluidic systems design process should, as in any other technology, proceed in an orderly manner. In general the major steps in proper sequence are the following:

1. Definition of system requirements.
2. Definition of component requirements.
3. Selection of components.
4. Analysis of performance.
5. Studies of critical parameters.

System requirements are basic, covering functional (what is the job to be done), environmental (what kinds of stresses will it be subjected to), and physical (where the components must be located) aspects.

Definition of component requirements involves system analysis to break down the overall requirements into specific requirements of each individual component.

Component selection applies the results of system analysis and the definition of system requirements to eliminate the obviously unsuitable components and to "zero in" on the optimum component for the application.

Performance analysis means putting together on paper all of the individual components of the system to determine if and how the combination will perform.

Finally parameter studies involve the determination of the effect of unwanted variables on system performance.

2.2 DEFINITION OF SYSTEM REQUIREMENTS

System requirements are primarily determined by the potential user with some inputs from the supplier depending on the user's familiarity with the fluidic technology. The user typically says:

1. I need a system to control a blank output in a certain prescribed manner and according to the behavior of certain inputs.
2. I have blank input which provides a signal of blank.
3. The system must operate from a blank psi line with blank contamination and consume no more than blank cfm.
4. It must operate for blank million cycles with blank maintenance in an environment of blank temperature and blank ambient pressure.
5. I'm evaluating this fluidic system against a blank system so it cannot cost more than blank, weigh no more than blank and occupy no more space than blank [and these last three blanks usually approach zero].

Then the potential supplier visits with the user to see first-hand what is to be done. He finds that the required system is to be operated by primates, maintained by cretins, and its failure means the end of the world. He also finds that the input is 300 ft from where the system can be mounted and the available location is where cutting oil is dripping on it constantly.

Fluidics people are concerned about these system requirements in various ways. First they ask, can fluidic devices perform the required functions? The answer in most cases is yes. Fluidic devices are available today to do nearly any kind of job imaginable. They can sense most physical variables (pressure, temperature, position, velocity), perform nearly every logic function (OR, AND, count, flip-flop, memory), provide all kinds of gain (proportional to 100,000), handle large quantities of power (up to hundreds of horsepower), and interface with other systems (high pressure pneumatic, hydraulic, and electric).

The second question about systems requirements is, can fluidic devices operate reliably in the environment? The answer is nearly always yes *providing the air supply is properly conditioned*. If this is the case, the fluidic system can be designed so the components are constantly purging themselves with clean, cooled, and dried air.

Fluidic systems are most attractive in explosive environments, where the cost and safety aspects of explosion-proof electric controls give a fluidic system a clear advantage. They are also most attractive in high temperature applications. They are less advantageous in moist and oily environments where condensed liquids can drip onto the devices and attract other forms of dust and dirt. In any case it is wise to enclose everything possible inside an enclosure like a NEMA 12 electrical box, keeping it slightly pressurized to keep dirt from entering.

The third system consideration is component layout. Size is sometimes a problem, and distance between components is nearly always a source of difficulty. Long interconnecting tubes cause significant pressure drops and even worse—time delay. The solution to this problem is to (a) use integrated circuits or manifolds whenever practical and (b) use booster amplifiers at the sending end of long lines.

2.3 COMPONENT REQUIREMENTS

System analysis to determine component requirements is the next step in the design process. One may start with a block diagram, isolating each function that must be performed between the input and output interface (see Figure 2.1). Or one can start with a functional diagram using the symbols shown in Figure 2.2, which are now standard in the technology. The system diagram would then take the form shown in Figure 2.3. Note that on both block and functional diagrams are shown the parameters available, such as line pressure, input movement, and output force. Now components must be found that can perform the functions indicated, match with the interfaces, and match with each other.

2.4 DESCRIPTION OF COMPONENTS

There are a number of ways to describe fluidic component characteristics but unfortunately no one method is accepted as standard by all manufacturers. The most widely used is, in general, graphical, but again the variables displayed are not consistent among manufacturers. Considerable direction toward a common description has been made by NFPA and SAE Standards Committees.

Figure 2.1 Block diagram of typical fluidic system.

34 The Fluidic Systems Design Process

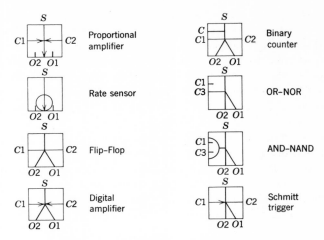

Figure 2.2 Standard functional fluidic symbols.

The basic information needed by the systems designer about a component's behavior is input and output characteristics. He also needs supply characteristics and the effect of external parameters. Because fluidic devices are inherently nonlinear the simplest way to describe these characteristics is by means of graphs. Unfortunately it is not practical to include time-dependent characteristics, so graphs are limited to displaying static characteristics.

Typical input characteristics of a vented jet-interaction proportional amplifier are shown in Figure 2.4. The characteristics are a plot of the flow into one control port versus the pressure applied at that port. The locus of bias points is the curve generated when both control port pressures are always made equal. The differential control curves result when one control port pressure is increased and the other decreased the same

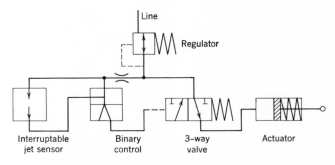

Figure 2.3 Functional diagram of typical fluidic system.

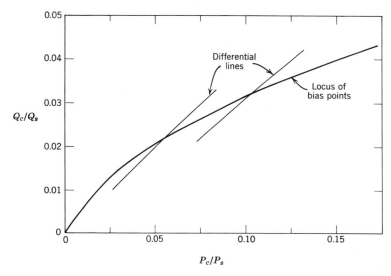

Figure 2.4 Input characteristics of jet-interaction amplifier.

amount, keeping the average of the two always at a fixed bias pressure. This is the way the amplifier would operate in a "push-pull" circuit.

The output characteristics shown in Figure 2.5 are a plot of the output pressure and flow as the load is varied from near zero resistance (high flow) to near infinite resistance (blocked output port). Because the output characteristics are also a function of the control signal, a complete description of the output characteristics is a family of curves of output flow versus output pressure, with control pressure or control flow as a parameter.

The characteristics of the wall-attachment flip-flop, a bistable device, can be shown with similar graphical characteristics. Input characteristics shown in Figure 2.6 define the relationship between control flow and control pressure at one port while the pressure at the other control port is held constant. Note the discontinuities at the switching points.

Figure 2.7 shows the output characteristics of the wall-attachment flip-flop. Since it has only two stable states, there are only two curves defining the relationship between pressure and flow as measured at the output port.

The foregoing input and output characteristics were normalized with respect to supply pressure and supply flow. This is a common practice that is reasonably valid and avoids the need for a set of curves at every allowable supply pressure. To interpret them, one more characteristic curve is required, the power nozzle characteristic, as shown in Figure 2.8. This is simply a plot of the pressure and flow into the supply port.

36 The Fluidic Systems Design Process

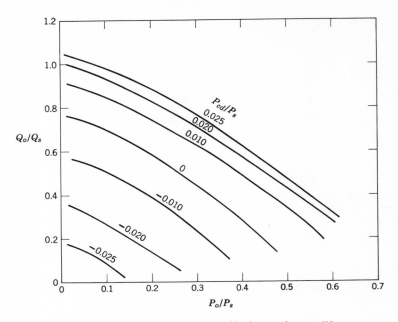

Figure 2.5 Output characteristics of jet-interaction amplifier.

As mentioned previously, graphical characteristics cover the static case only. Dynamic time-dependent characteristics must be handled another way. Commonly they are given simply as "frequency response, 1000 hertz" or "10 milliseconds into a one cubic inch load." These descriptions are very ambiguous and incomplete. A better way is by equivalent electrical circuits. The technique has been used widely in other

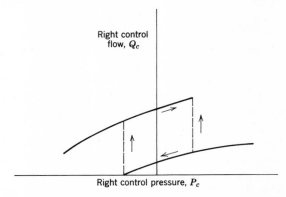

Figure 2.6 Input characteristics of wall-attachment amplifier.

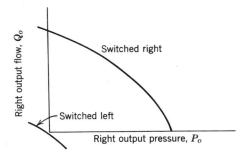

Figure 2.7 Output characteristics of wall-attachment amplifier.

technologies to describe anything from mechanical vibrations to the human vocal tract. Although they are valid for small changes only, they have advantages of being simple to cascade, compatible with well-known analytical tools, and able to provide tremendous insight into system operation.

A typical equivalent circuit is illustrated for the jet-interaction proportional amplifier in Figure 2.9. The element in series with the input circuit $2L_c$ is due to inertance in the line to the control nozzle. The shunt elements $2R_c$ and $C_c/2$ are effective control nozzle resistance and volume capacitance of the control aperture. The equivalent generator $2K_p$ contains a delay factor (e^{-st_d}) that includes wave propagation and transit times in the total path from the control port to the load terminals. The output circuit contains a series inductor $2L_o$ and restrictor $2R_o$ and a shunt

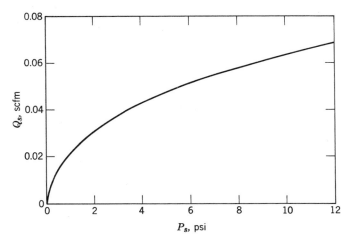

Figure 2.8 Supply port characteristics of jet amplifier.

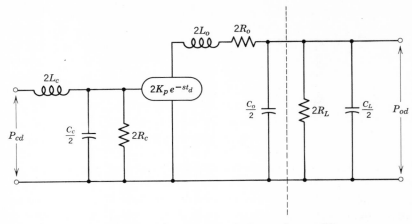

Figure 2.9 Equivalent circuit for jet-interaction amplifier.

volume capacitance of $C_o/2$. If the lines to the load are short, the load volume capacitance $C_L/2$ is directly in parallel with the amplifier capacitance, and the load resistance $2R_L$ parallels both.

The elements that appear in the electrical equivalent circuits can all be calculated from the graphical characteristics, circuit dimensions, and conditions at the bias (quiescent, no-signal) operating point.

A third way to describe fluidic components is by means of matrices. Matrix describers account for all cross-coupling terms such as internal feedback and coupling between control and supply ports. A typical admittance matrix describer in general terms is illustrated in Figure 2.10. The first row accounts for conditions at one control port including input flow, feedback flow from the output ports (2), cross flow from the opposite control port, coupling with the supply port and coupling with the vents (2). There is a separate row (equation) describing the conditions at each port.

$$\begin{bmatrix} w_1 \\ w_2 \\ w_3 \\ w_4 \\ w_5 \\ w_6 \\ w_7 \end{bmatrix} = \begin{bmatrix} Y_{11} & Y_{12} & Y_{13} & Y_{14} & Y_{15} & Y_{16} & Y_{17} \\ Y_{21} & Y_{22} & Y_{23} & Y_{24} & Y_{25} & Y_{26} & Y_{27} \\ Y_{31} & Y_{32} & Y_{33} & Y_{34} & Y_{35} & Y_{36} & Y_{37} \\ Y_{41} & Y_{42} & Y_{43} & Y_{44} & Y_{45} & Y_{46} & Y_{47} \\ Y_{51} & Y_{52} & Y_{53} & Y_{54} & Y_{55} & Y_{56} & Y_{57} \\ Y_{61} & Y_{62} & Y_{63} & Y_{64} & Y_{65} & Y_{66} & Y_{67} \\ Y_{71} & Y_{72} & Y_{73} & Y_{74} & Y_{75} & Y_{76} & Y_{77} \end{bmatrix} \begin{bmatrix} p_1 \\ p_2 \\ p_3 \\ p_4 \\ p_5 \\ p_6 \\ p_7 \end{bmatrix}$$

Flow = Admittance times pressure

Figure 2.10 Matrix description of fluidic amplifier.

It is clear that the matrix describer is one step more sophisticated than the equivalent circuit representation, although it normally has the same capabilities and limitations. It takes into account all coupling terms but it is normally limited to small-signal changes. Using electronic computers for solving the equations, it could be feasible to use nonlinear coefficients, in which case both static and dynamic problems could be handled.

The admittance matrix method for describing fluidic devices is just in its infancy. The basic approach has been layed out but so far the job of measuring coefficients looking into each port has not been thoroughly worked out. However the method is the most powerful so it is just a matter of time before the measurement problems are solved and fluidic devices can be described conveniently by matrices.

There is a fourth way to describe fluidic component characteristics and that is as a continuous physical system. This is the approach originally taken by those trained in fluid mechanics. They wrote all of the equations describing the fluid system inside a component, then tried to solve the resulting array. Some significant progress has been reported recently but the problem is so complex it is not suitable for practical application in systems design.

2.5 COMPONENT SELECTION

The next step in the systems design process is the selection of components to meet the requirements previously defined.

The most important requirement of a component is that it performs the proper function. Therefore we would first assemble catalog information by function such as "OR" gates or proportional amplifiers. One of the most important criteria is pressure level, which must be compatible throughout the system. Impedance levels must also be compatible although in some cases buffer devices can be inserted between mismatched stages. Speed of response is another of the most important but only in certain applications. If efficiency is important, air consumption and recovery factors must also be considered. Environment may be critical so components must be examined for sensitivity to contamination, vibration, and acoustic noise. Finally although we may hesitate to admit it, company loyalties play an important part in component selection, which often overrides selection on characteristics alone.

The outstanding features of the various types of fluidic components currently available will be covered briefly in Chapter 3. It is intended to point out generally the kinds of things to look for in selecting components.

40 The Fluidic Systems Design Process

2.6 CALCULATION OF PERFORMANCE

Just as there are four different ways to describe fluidic components, there are four ways to analyze system performance using these component descriptions.

Graphical characteristics are used in system design by applying the "load line" technique as illustrated in Figure 2.11. The output characteristics of the driving component and the input characteristics of the driven component are then superimposed. The intersections of the two sets can be the only stable operating points.

Since the family of curves making up the output characteristics have the control variable of the driving component as the parameter, the transfer curve of the coupled components can be determined by plotting the output pressure against the control variable at each point of intersection.

Equivalent electric circuits are used directly in the dynamic analysis of fluidic systems as illustrated in Figure 2.12. The basic equivalent circuits of each component are determined. The equivalent circuits are then interconnected in a manner analogous to the actual fluidic system. The circuit is then analyzed using conventional mathematical techniques. The transfer function is determined from a simultaneous solution of all the loop

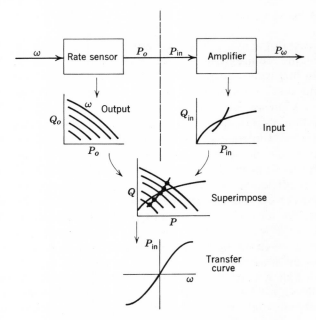

Figure 2.11 Graphical system analysis.

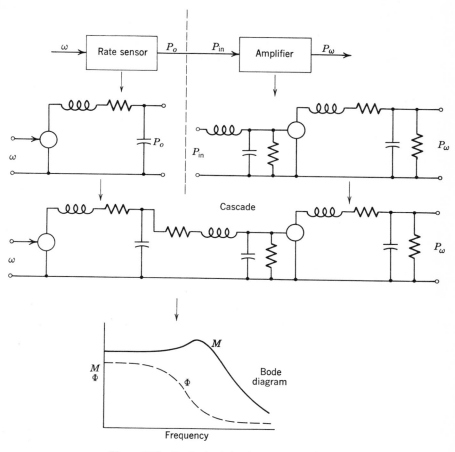

Figure 2.12 Equivalent circuit systems analysis.

equations. At that point frequency can be entered as the variable to calculate the frequency response and plot it as a Bode diagram.

The matrix describers are used in systems analysis according to certain simple rules. For example, as illustrated in Figure 2.13, the matrices describing the coupled components would first be determined. According to the number of ports connected between the two components we would set the pressures equal and the sum of the flows zero. Thus, we can generate a matrix describing the interconnected components. For simple stability analysis, the admittance matrices can be used directly.

The direct mathematical description of fluidic components and systems is in the early stages of development. For this approach, component

42 The Fluidic Systems Design Process

$$\omega \rightarrow \boxed{\begin{array}{c}\text{Rate}\\\text{sensor}\end{array}} \xrightarrow{p_3, p_4 \quad w_1, w_2} \boxed{\text{Amplifier}} \xrightarrow{p_3, p_4}$$

$$\downarrow A \qquad\qquad\qquad \downarrow B$$

$$\begin{bmatrix}0\\0\\w_3\\w_4\end{bmatrix} = \begin{bmatrix}0 & 0 & Y_{13} & Y_{14}\\0 & 0 & 0 & 0\\0 & 0 & Y_{33} & Y_{34}\\0 & 0 & Y_{43} & Y_{44}\end{bmatrix}\begin{bmatrix}\omega\\0\\p_3\\p_4\end{bmatrix} \qquad \begin{bmatrix}w_1\\w_2\\w_3\\w_4\end{bmatrix} = \begin{bmatrix}Y_{11} & Y_{12} & Y_{13} & Y_{14}\\Y_{21} & Y_{22} & Y_{23} & Y_{24}\\Y_{31} & Y_{32} & Y_{33} & Y_{34}\\Y_{41} & Y_{42} & Y_{43} & Y_{44}\end{bmatrix}\begin{bmatrix}p_1\\p_2\\p_3\\p_4\end{bmatrix}$$

$$\begin{array}{c}\underline{w_a} + \underline{w_b} = 0\\\text{Partially } \underline{p_a} = \underline{p_b} = \underline{p}\end{array} \quad \text{Interconnection}$$

$$\begin{array}{c}(\underline{Y_a} + \underline{Y_b})\underline{p} = 0 \quad \text{Stability}\\\text{and det. } (\underline{Y_a} + \underline{Y_b}) = 0 \quad \text{analysis}\end{array}$$

Figure 2.13 Matrix systems analysis.

description is rather inseparable from systems analysis. Consequently, the difficulty of the problem is compounded by the need to consider both at the same time. The behavior of the fluid throughout the entire system must be described mathematically by numerous coupled nonlinear expressions that involve the practically insoluble Navier–Stokes equation. This solution will eventually be the ultimate in accuracy, accounting for all the nonlinearities as well as all secondary variables. However, it will be many years before this approach will become practical, in spite of the capabilities of the electronic computer.

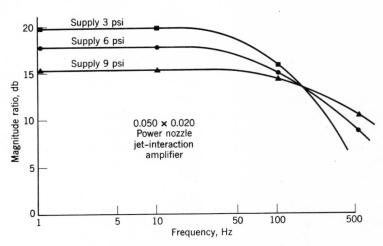

Figure 2.14 Effect of supply pressure on frequency response.

2.7 PARAMETER STUDIES

In concluding the design of a critical system it is often wise to investigate the effect of parameter changes. One of the most important is supply pressure, which can compromise performance to the point of complete failure. A typical case would be the effect of supply pressure on frequency response. The results calculated and tested for a single stage amplifier are shown in Figure 2.14. Note that there is a significant shift in the natural frequency as well as an expected change in gain.

A second important type of parameter study is to determine failure characteristics. If the system is doing a critical job it would be necessary to make it fail-safe and to indicate its own failure.

Other parameters that might be the object of study are temperature, length of tubing, and changes in ambient pressure.

2.8 SUMMARY OF THE DESIGN PROBLEM

In summary, so far we have outlined the procedure for systematic design of fluidic systems. It covers the following major steps.

1. Definition of system requirements.
2. Definition of component requirements.
3. Selection of components.
4. Analysis of performance.
5. Studies of critical parameters.

3
Operating Principles of Fluidic Devices

3.1 ANALOG AMPLIFIERS

Vented Jet-Interaction Amplifier

The simplest configuration of fluid amplifier is the vented jet-interaction amplifier, illustrated in Figure 3.1. Pressure is applied at the supply port to produce a high-velocity jet issuing into the vented interaction region. Because this region is vented, essentially the power stream is issuing into an area maintained at relatively constant pressure, with very little interaction with the surrounding walls. If it is not deflected, the center line of the power stream will strike the splitter. Half of the power stream exits out the right output port and the other half of the power stream exits through the left output port. If equal impedances are placed downstream of the output legs and the pressure is measured between the output ports, the differential output is zero.

Now if a pressure difference is applied between control ports, the resulting momentum flow will deflect the power stream, and if the higher pressure is applied to the right control port the stream will be deflected so that more flow goes out the left output port than the right output port. Then the pressure difference measured between output ports is a finite value.

The angle of deflection of the power stream is proportional to the differential pressures and flows applied between control ports and the power stream can be deflected until 100% goes out the left output port and none goes out the right output port. The power stream can be deflected to the right output port by applying a higher pressure on the left control port

Analog Amplifiers   45

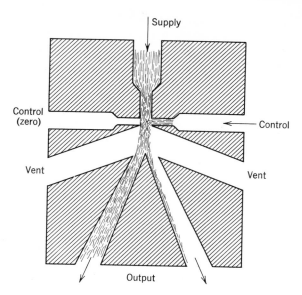

Figure 3.1 Typical vented jet-interaction amplifier.

than on the right control port. A typical gain curve for a vented jet-interaction amplifier is shown in Figure 3.2.

In general, vents are provided in proportional jet-interaction amplifiers to minimize the effect of loading; that is, the effect of partially blocking the output ports. Since the vented interaction region is always at a relatively constant pressure, the pressures that back up in the output legs have

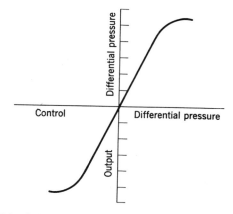

Figure 3.2 Gain curve of typical vented jet-interaction amplifier.

little effect on the interaction between control and power streams. In addition, as ports become blocked, the vents will provide an escape path for the excess fluid in the power stream.

Parameters of importance in the design of proportional jet-interaction amplifiers are the following:

1. Aspect ratio of the jets.
2. Relative and absolute sizes of control and power nozzles.
3. Setback of control nozzles from power stream.
4. Length of interaction chamber.

Important features of the vented jet-interaction amplifier are the following:

1. High pressure gain (5).
2. Relatively high frequency response (KHz).
3. Medium input impedance.
4. Medium output impedance.
5. Low efficiency.
6. Medium signal to noise ratio.
7. Stable under most loads.

Closed Jet-Interaction Amplifier

The closed jet-interaction amplifier is illustrated in Figure 3.3. Again pressure is applied to the supply port to produce a power jet that issues from the power nozzle into the closed interaction region. If the power stream is undeflected, its center line will strike the splitter and half the flow will go out the right output port and half the flow will go out the left output port, producing a zero differential pressure across the output ports. When a pressure difference is applied across the control ports the power stream is deflected, causing more flow to exit through one output port than through the other output port, producing an output differential pressure.

The deflection of the power stream is proportional to the differential pressures applied at the control ports, so that the differential output pressure is also proportional to the differential control pressures applied. A typical gain curve for an amplifier of the closed jet-interaction type is illustrated in Figure 3.4.

Without the vents to atmosphere, the effect of loading the output ports, that is, restricting the output flow, is reflected back to the interaction region, affecting the interaction between the power stream and the control streams. When the output ports are restricted, the pressure backs up

Analog Amplifiers   47

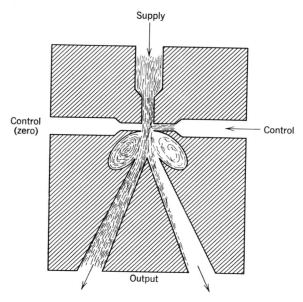

Figure 3.3   Typical closed jet-interaction amplifier.

into the interaction region raising the entire internal pressure of the device. Since the power jet is issuing into a higher pressure, the pressure difference driving the power jet is less and the strength of the power stream is decreased. In the extreme case when the output ports are totally blocked, the power jet flow becomes negligible and the amplifier is inoperative.

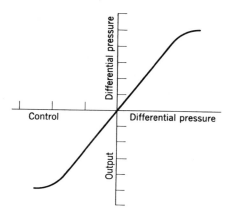

Figure 3.4   Gain curve of typical closed jet-interaction amplifier.

48 Operating Principles of Fluidic Devices

Features of the closed jet interaction amplifier are similar to those of the vented type except for (a) medium pressure gain and (b) inoperative with blocked loads.

Vortex Amplifier

The vortex amplifier, in various forms, has been developed by a number of companies. The basic vortex amplifier is illustrated in Figure 3.5. Pressure is applied to the radial inlet port, producing a power jet that issues from the radial nozzle into a short cylindrical chamber. If the power stream is undeflected it will continue toward the center of the chamber and exit through an outlet port maintained at a lower pressure or returned to ambient through a load restrictor. A control pressure is applied to the tangential inlet port producing control flow from the tangential nozzle. This deflects the power stream and imparts a component of momentum to

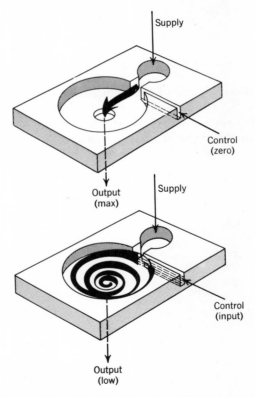

Figure 3.5 Typical vortex amplifier.

the tangential direction. As the stream flows toward the outlet port, it forms a vortex swirling at higher and higher velocity until it goes out the outlet port. As a result of the vortex, the pressure drop across the vortex chamber is increased in proportion to the tangential momentum injected. With an impedance connected to the output port the pressure at the output port is inversely proportional to the pressure applied to the tangential input port.

A typical gain curve is illustrated in Figure 3.6. The vortex device illustrated here functions primarily as a throttling valve, varying the impedance between the supply and the output ports. As such it is unique in fluidics, being the only proven device with the capability for a reasonable "turn-down" of the flow from the power source. Other vortex amplifiers employ a receiver that is shaped to accept or reject the cone-shaped flow as it exits from the vortex chamber.

Parameters of importance in the design of vortex amplifiers and valves are the following:

1. Ratio of chamber to outlet diameters.
2. Aspect ratio of the vortex chamber.
3. Uniformity of supply flow into the chamber.
4. Location of the control nozzles.

Important features of the vortex devices are the following:

1. Medium pressure gain.
2. Low frequency response.
3. "Turn-down" capability (as high as 10:1).

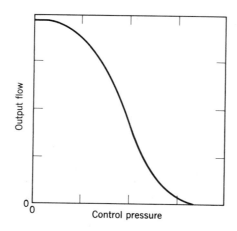

Figure 3.6 Typical vortex amplifier control curve.

4. Medium input impedance.
5. Medium output impedance.
6. Medium noise.
7. Stable under all loads.

Boundary-Layer-Control Amplifier

The boundary-layer-control amplifier is shown in two of its simplest forms in Figure 3.7. Various configurations have been developed. Pressure is applied to the supply port to form a power stream that issues over a curved surface. The power stream tends to remain attached to the curved surface for a considerable distance downstream. When the proper conditions have been satisfied, this stream separates from the curved surface, strikes a conventional splitter, and exits through one of two output ports.

A control stream is injected into the boundary layer where the power

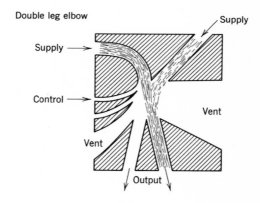

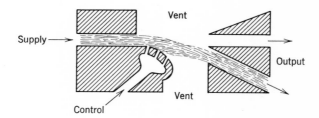

Figure 3.7 Typical boundary-layer-control amplifier.

stream is attached to the curved surface. The injected flow tends to separate the power stream sooner, so it is directed at a different angle and more flow exits through one output port than through the other. The point of separation and thus the angle of deflection is proportional to the amount of flow injected into the boundary layer. Therefore with impedances connected to the output ports, the differential pressure measured between output ports is proportional to the pressure applied to the control port.

A typical gain curve for the double leg elbow amplifier is illustrated in Figure 3.8. Note that the input is "single-ended" but the output can be differential.

Vents are usually provided to avoid the effects of output port loading on the performance of the amplifier and to provide a path for the rejection of excess fluid.

Parameters important in the design of boundary-layer-control amplifiers are the following:

1. Aspect ratio of the power stream.
2. Location of the auxiliary parts (splitter, etc.).
3. Velocity of the power stream.
4. Curvature of the control surface.
5. Number and location of the control slots.
6. Directivity of the control slots.
7. Length of the interaction chamber.

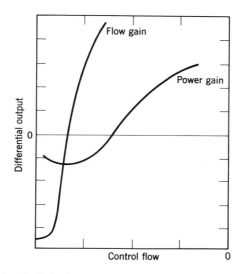

Figure 3.8 Typical gain characteristics of double leg elbow amplifier.

Important features of the boundary-layer-control amplifier are the following:

1. High flow gain (200).
2. Medium frequency response.
3. Medium input impedance.
4. Low output impedance.
5. Broad low-velocity streams (low frequency noise).
6. Low pressure output.
7. High efficiency.
8. Stable under all loads.

Impact Modulator

The impact modulator is illustrated in Figure 3.9. Supply pressure is applied to two input ducts producing two jets that strike head-on in the interaction region. (Note that the design is symmetrical around the center line through the input ducts so the jets are circular.) The relative strengths of the two power jets are adjusted so the effective point of impact is just outside the output duct. Under these conditions the output flow is negligible and the device is effectively turned off.

When a control pressure is applied to the control duct a flow is added to the stream issuing from the left side. This causes the effective point of impact to move inside the output duct, producing an output flow. The distance that the point of impact moves inside the output duct is proportional to the amount of control pressure and flow; therefore, the output flow is proportional to the control pressure and flow. If an impedance is connected at the output port, the output pressure is proportional to the control pressure. A typical gain curve is shown in Figure 3.10. The effect of loading at the output port is normal and the amplifier is stable with a wide range of loads.

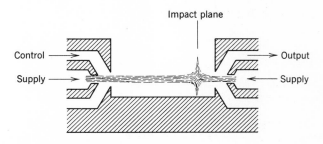

Figure 3.9 Typical direct impact modulator.

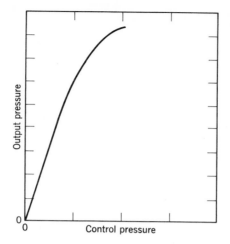

Figure 3.10 Typical impact modulator control curve.

The parameters of importance to the design of the impact modulators are as follows:

1. Nozzle separation to diameter ratio.
2. Relative pressures on opposing power nozzles.
3. Relative diameters of power, control, and output ducts.
4. Method of introducing control (direct or transverse).

The important features of the impact modulator are as follows:

1. High pressure gain.
2. High input impedance.
3. Output resistance variable with supply pressures.
4. No internal feedback.
5. Medium frequency response.
6. High signal-to-noise ratio.
7. Stable with all loads.

3.2 ANALOG SENSORS

Back-Pressure Nozzle

The most common type of sensor used in the fluidics field is the sensing jet. It is based on the principles of pneumatic gages and flapper-nozzle devices that have been in use for many years.

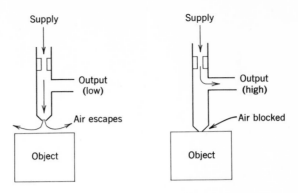

Figure 3.11 Back-pressure nozzle.

The sensing jet is illustrated in Figure 3.11. Pressure is applied to a network made up of a fixed restrictor in series with the sensing nozzle. The output pressure is taken at the junction of the two. When the nozzle is uncovered, its impedance is relatively low, so the major portion of the pressure drop occurs across the fixed restrictor. Therefore the output pressure measured is relatively low. When the jet is partially covered, the impedance of the nozzle is increased a significant amount, so the major portion of the pressure drop occurs across the nozzle rather than across the fixed restrictor. In this case the output pressure is relatively high.

This arrangement is often used to sense the presence or absence of an object. The output pressure is proportional to the degree the nozzle is covered, so the sensing jet can be used as a sensor in either analog or digital systems. With an appropriate coding disk or plate, it can be used as a position encoder or as a pulse generator.

Interruptable Jet

The interruptable jet, commonly known as the "fluidic eye," has been developed by a number of device manufacturers as an object sensor. It is directly analogous to the electronic photocell or electric eye, hence its common name.

The interruptable jet is illustrated in Figure 3.12. A supply pressure is applied to the input tube to produce a persistent turbulent jet, which is projected several nozzle diameters toward a receiver. When there is no interference a high percentage of the nozzle pressure is recovered in the receiver, just as in the jet-interaction amplifier. When the projected jet is interrupted, little or no pressure is recovered in the receiver. Therefore, the pressure in the receiver is an indication of the presence of an object

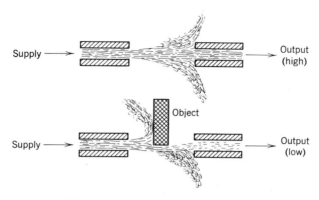

Figure 3.12 Turbulent interruptable jet.

between the nozzle and the receiver. The pressure in the receiver is proportional to the degree the jet is interrupted, so the output (receiver) pressure is proportional to the position of the interrupting object. Thus the interruptable jet can be used as an analog sensor as well as a digital sensor.

Bubbler Tubes

For sensing the level of liquids and certain dry bulk materials, a simple arrangement of restrictors and tubing is often used. As shown in Figure 3.13 supply air is applied to a tube through a calibrated orifice. An output signal is generated at their juncture. When the tube is uncovered, air bleeds out and the output signal is low. When the tube is submerged and bubbles are forced from its end the pressure at the output is proportional to the "head." For example, if water rises 2.76 in. above the end of the tube, the output pressure is exactly 0.1 psi.

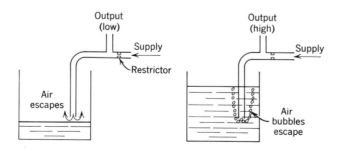

Figure 3.13 Bubbler tube sensor.

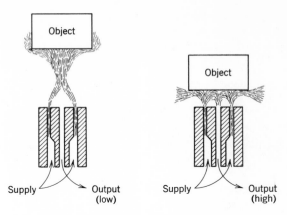

Figure 3.14 Converging jet sensor.

Converging Jet Sensor

To overcome the air consumption limitation of the back pressure sensor, the converging cone sensor uses an annular jet concentric with an output passage. As shown in Figure 3.14 the high-velocity annular jet, after leaving the nozzle, quickly converges, containing an area of low pressure inside the bubble formed. The output signal is therefore slightly below atmosphere.

As a target moves into range some of the jet flow is reflected into the

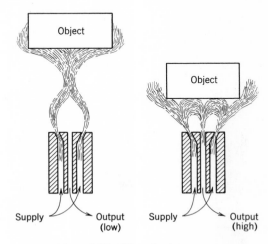

Figure 3.15 Diverging jet sensor.

low pressure region. Therefore the output signal pressure is increased an amount inversely proportional to the distance between target and sensor. Range is nearly one nozzle diameter.

Diverging Jet Sensor

To further increase the sensing range without more air consumption, the annular jet is directed at an angle away from the nozzle centerline (see Figure 3.15). Now the air is projected a greater distance before joining, creating a longer bubble and a lower pressure inside the bubble. Therefore the flow begins to be reflected into the bubble at a greater distance between nozzle and target, hence a longer sensing range.

Diverging cone sensors have a range up to two nozzle diameters and are practical up to 1 in. or more. Their output is inversely proportional to the distance to a target.

Vortex Proximity Sensor

Object sensors based on the vortex principle generally operate on the back pressure generated inside a vortex chamber as illustrated in Figure 3.16. As the flow exits from the chamber it is swirling rapidly. This creates a diverging cone pattern that is normally entraining ambient air into its center.

The output is taken at the center of the vortex inside the chamber. When there is no target in view, the divergent cone is open and the center of the vortex is at ambient pressure. When a target approaches it blocks

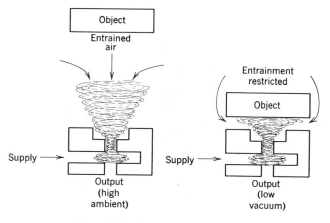

Figure 3.16 Vortex proximity sensor.

the entrainment of air into the center of the vortex and a vacuum is generated there. The phenomenon is analog in nature so the output vacuum signal is inversely proportional to the distance to the target. Maximum range is 0.2 in.

Vortex Rate Sensor

One of the best known analog fluidic sensors is the vortex rate sensor. It has been developed by numerous agencies, including Harry Diamond Laboratories.

The vortex rate sensor is illustrated in Figure 3.17. Supply pressure is applied to a manifold surrounding the cylindrical vortex chamber. Uniform flow enters the chamber through a porous wall. With no disturbances the flow proceeds uniformly straight toward the drains located in the center of the cylindrical vortex chamber and exits in a relatively laminar fashion through the drain tubes.

When the body of the rate sensor is turned at a uniform rate about the cylinder axis, the fluid flowing through the porous wall receives a component of momentum tangential to the vortex chamber. Then as the flow proceeds toward the drain, it forms a vortex of increasing angular velocity until it exits through the drains with a motion forming a helix. The angle

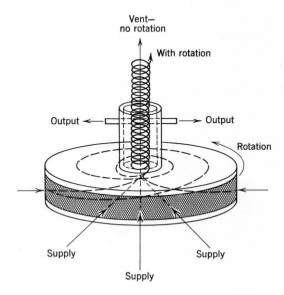

Figure 3.17 Typical vortex rate sensor.

of the helix is proportional to the momentum imparted as the fluid flows through the porous wall and is therefore proportional to the rate of turn.

The most common method for providing an output from the vortex rate sensor is by means of pickup tubes inserted across the drain. Small holes are drilled in the walls of the tubes on either side of the drain centerline and a barrier is provided inside the tubes between the holes. When the flow is issuing out the drain in a laminar fashion both pickup tubes sense the same pressure. When the flow is in the form of a helix, one tube senses a higher pressure and the other tube senses a lower pressure. The pressure difference is proportional to the angle of the helix and its direction. Therefore, the differential pressure output is directly proportional to the magnitude and direction of the rate of turn of the body of the vortex rate sensor. A typical sensitivity curve is illustrated in Figure 3.18.

3.3 DIGITAL AMPLIFIERS

Wall-Attachment Amplifier

The fluidic device that was developed first was the wall-attachment amplifier. The wall-attachment amplifier is illustrated in Figure 3.19. Pressure applied at the supply port produces a power jet that issues from the power nozzle into the interaction region. Initially some of the flow goes out the left output port and some of the flow goes out the right output port. However, this is an unstable condition. Due to the entrainment produced by the power jet as it issues from the nozzle, there is a low pressure region generated in the vicinity of the two walls. The jet is attracted to one or the

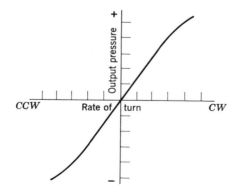

Figure 3.18 Typical vortex rate sensor sensitivity curve.

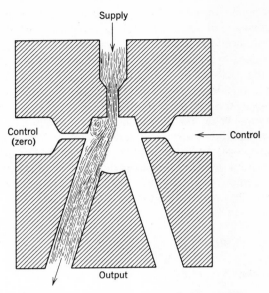

Figure 3.19 Typical nonvented wall-attachment amplifier.

other and as it deflects closer and closer to the wall of its choice, the pressure becomes increasingly low until finally the power stream attaches to that wall. This is known as the Coanda effect.

Now if the control flow is injected into the low pressure region, the pressure there is increased, permitting the power stream to detach from that wall. Because there is still a low pressure region due to entrainment on the opposite side, the power stream is attracted to the opposite wall, switching over and attaching to that side.

The procedure can be repeated by decreasing the pressure on the right control port and increasing the pressure on the left control port, once again switching the power stream to reattach to the right wall. With impedances connected to the output ports there is a pressure difference generated between output ports that is a function of the switched position of the power jet. Therefore, we have an output pressure that is a function of the control pressures and the memory of the device. This is equivalent to a flip-flop function. A typical switching characteristic is illustrated in Figure 3.20.

Vents are often provided in the wall-attachment amplifier to avoid the feedback effect of loading. However, they cannot be placed to vent the interaction region or there would be no attachment. Therefore, they must be placed downstream in the output legs.

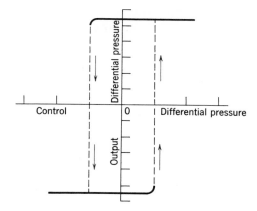

Figure 3.20 Typical switching characteristic of nonvented wall-attachment amplifier.

Other design parameters of importance are the following:
1. Aspect ratio of the jets.
2. Relative and absolute sizes of control and power nozzles.
3. Setback of control nozzles from power stream.
4. Length and setback of attachment wall.
5. Length of interaction chamber.
6. Shape of splitter.
7. Position of vents.

Important features are as follows:
1. High switching sensitivity.
2. Good memory.
3. Relatively fast response.
4. Medium input impedance.
5. Medium output impedance.
6. Medium signal to noise ratio.
7. Low efficiency.
8. Stability under most loads.

Turbulence Amplifier

The turbulence amplifier is one of the most practical and widely used fluidic devices. It was invented by Raymond Auger.

The turbulence amplifier is illustrated in Figure 3.21. Pressure is applied to the supply tube, producing a low-velocity jet issuing into the interacion

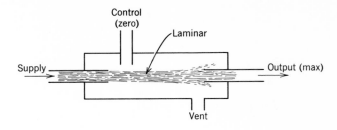
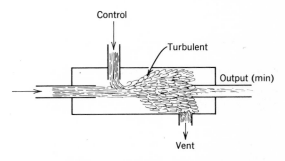

Figure 3.21 Typical turbulence amplifier.

chamber. Conditions are chosen so that the jet is perfectly laminar to a receiver just inside where it would normally become turbulent. Most of the flow is recovered and flows out the output tube. When a control pressure is applied to any one of the control tubes, the laminar flow is disturbed at a point ahead of where the flow enters the output tube; thus, the amount of flow recovered in the output tube is decreased. The amount the output is decreased is proportional to the control flow, but it is highly sensitive phenomena so the turbulence amplifier is used only as a digital, that is an on-off device. The interaction chamber is vented to avoid the effects of loading of the output tubes.

A typical control characteristic is illustrated in Figure 3.22.

Axisymmetric Focused-Jet Amplifier

The axisymmetric focused-jet amplifier is essentially a three-dimensional wall-attachment amplifier and is therefore used as a digital device.

The axisymmetric focused-jet amplifier is illustrated in Figure 3.23. Supply pressure is applied to the input duct providing a uniform pressure in the manifold or power chamber. This produces an axisymmetric sheet of fluid that attaches to the walls of the drain directing it out the output

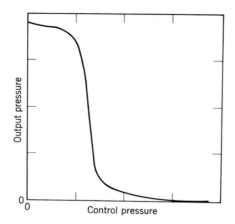

Figure 3.22 Typical turbulence amplifier control characteristic.

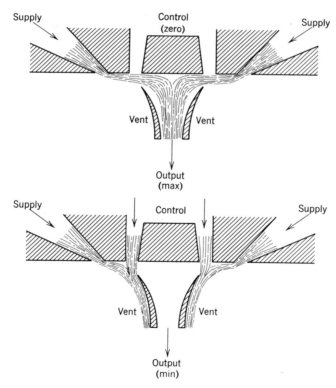

Figure 3.23 Typical axisymmetric focused-jet amplifier.

64   Operating Principles of Fluidic Devices

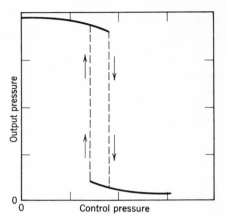

Figure 3.24   Typical axisymmetric focused-jet amplifier switching characteristics.

ports. Pressure applied to the control ducts produces a flow that is injected into the boundary layer to separate the flow from the walls of the output drain. This causes the sheet of fluid to open up, decreasing the amount of flow out the output ducts. Again the interaction region is vented to prevent the effects of loading on the phenomena occuring in the interaction region. A typical control characteristic is shown in Figure 3.24.

Passive Logic Devices

The wall-attachment amplifier can be designed to perform many digital logic functions such as inversion, OR-NOR, NOR, and AND/NAND. These devices provide both memory (latching) and power gain. However, in many cases the logic functions can be performed acceptably with passive devices, much less complex devices operating on the signal power alone.

A typical passive AND device is illustrated in Figure 3.25. Pressure signals are applied at both input ports, producing flows through the nozzles that intersect in the vented interaction region. The output receiver is placed downstream, on a line that bisects the angle between the two signal jets.

When there is no signal on either of the input ports there are, of course, no jets and no output. If a signal pressure is applied to *either* of the two input ports, a jet is produced, but because of the geometry, it misses the output receiver and exits through the vent, again no output.

If signal pressures are applied to *both* input ports, two jets of equal magnitude are produced that intersect in the interaction region. The

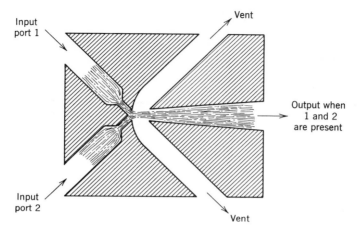

Figure 3.25 Typical passive AND device.

combined jet flows along the line that bisects the angle between the two intersecting jets and exits through the output ports. With an impedance downstream, an output pressure signal is produced. Thus we have a device that only produces an output signal when *two* input signals are present and no output signal when either one or both input signals are not present. In other words an output signal is produced only when input 1 and input 2 are present, an AND function.

The vents are provided to permit the escape of excess fluid and to minimize the effect of output port loading.

Important design parameters of the passive logic devices are rather obvious. Except for aspect ratio of the jets, the only important thing is the geometry necessary to perform the desired function.

Features of passive logic devices are as follows:

1. Simplicity.
2. High output impedance.
3. Low power output.

3.4 DIGITAL SENSORS

Limit Valves

Although they may be considered conventional pneumatic devices, various types of small valves are widely used with fluidic logic and control circuits. Some have a plunger-type extension connected to a spool or poppet,

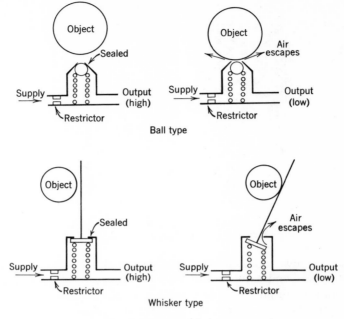

Figure 3.26 Limit valve sensors.

which can be depressed by a passing object, exactly as an electrical limit switch. Others, such as illustrated in Figure 3.26, use a spring-loaded bleed poppet. The ball operator, when depressed, allows supply air to bleed out, reducing the pressure at the output port. The whisker-type, when deflected, opens one side of a simple bleed poppet.

Limit valve sensors are most often used as digital sensors, performing the function of the common electrical limit switch. However, they can also be used as proportional sensors and are capable of measuring the motion of a target object with a resolution of better than 0.001 in.

Interrupted Laminar Jet

The interrupted jet can be operated in either one of two different modes. With a turbulent jet it is a proportional sensing device, as previously described. With a laminar jet it is bistable and sensitive to objects and motions of a few thousandths of an inch. As illustrated in Figure 3.27, a jet is projected at very low velocity, forming a coherent laminar stream across a gap to a receiver, where nearly all the flow is recovered. Then the pressure at the output is high. If an object interrupts the stream it becomes

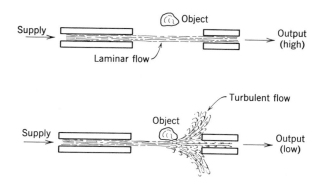

Figure 3.27 Laminar interrupted jet.

turbulent at the point of interruption and the stream is dissipated before it reaches the receiver. The output pressure becomes very low.

Diverging Jet Sensor

The diverging jet sensor is illustrated in Figure 3.28. A low pressure supply of air is connected to the supply port, forcing air out the annular nozzle. As the flow issues from the nozzle, all parts entrain air from the surrounding atmosphere. The divergent conical stream becomes convergent and attaches to itself, forming a closed bubble. Inside the bubble is a vacuum, holding it closed. A hole in the side of the nozzle conducts

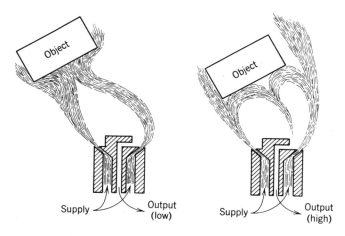

Figure 3.28 Long-range diverging jet sensor.

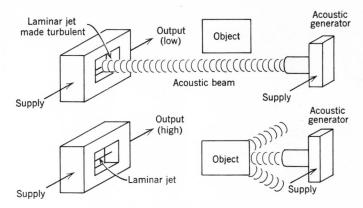

Figure 3.29 Acoustic beam sensor.

this vacuum to the signal port. Therefore when the bubble is undisturbed there is a vacuum signal.

If an object comes into the field of view of the sensor, which can extend for inches, the closed bubble bursts open and the signal pressure goes to atmospheric. Because this phenomenon is bistable, this diverging jet sensor is a proximity switch.

The outstanding features of the diverging jet proximity switch are as follows:

1. Long range (4 in. or more).
2. Requirement of access to only one side of target.
3. Bistable switch.
4. Fast response (3 to 4 msec).
5. Low vacuum (or pressure) output.
6. High output impedance.
7. Sensitivity to solid, liquid, porous, and transparent objects.

Acoustic Beam Sensors

Another fluidic equivalent of the photoelectric eye is the acoustic beam sensor (fluidic ear). As illustrated in Figure 3.29, a fluidic oscillator generates a sonic wave (50 kHz), which is focused and projected as a sonic beam. The receiver is usually a form of interrupted jet operated in the laminar mode. When the sonic beam impinges on the laminar jet, the beam causes it to become turbulent, reducing the output signal. When an object interrupts the sonic beam, the laminar jet is re-established and the output signal increases.

Acoustic sensors have ranges to 5 ft or more.

4
Component Description

4.1 GRAPHICAL CHARACTERISTICS

In designing any system of interconnected components it is necessary to take into account the effect of one upon the other. This is true whether the components are electronic, hydraulic, mechanical, acoustic, or fluidic.

The most practical way of doing this is the so-called "black box" method, which is widely used in control system design. This technique requires that each component be isolated from all other components in the system, then subjected to a few simple tests under typical operating conditions. This is normally done by the manufacturer. For example, the vacuum tube manufacturer supplies a set of characteristic curves and dynamic parameters for each tube type he markets.

The information is used in electronics for amplifier stage coupling (matching) and in hydraulics for servovalve and load coupling.

By applying the same proven approach to fluidics, all the mathematical tools now used in electronics and hydraulics can be applied to fluidic systems analysis and design.

For most practical cases it is possible to describe the total behavior of any fluidic device with the three sets of data shown in Figure 4.1. *Input characteristics* define what load an input signal sees when it is applied to the input ports. *Transfer characteristics* define what happens to the output when an input signal is applied. *Output characteristics* define how the output signal is affected when an external load is connected at the output ports.

For *static and large signal analysis* these three characteristics are most conveniently described graphically, because the graphs take into account

70   Component Description

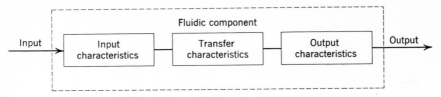

Figure 4.1 Characteristics of any fluidic component.

device nonlinearities without complex mathematics. For *dynamic and small-signal analysis* these characteristics are more conveniently described in terms of equivalent electrical circuits, because well-developed linear circuit theory is directly applicable to the calculation of performance.

Analog Amplifiers

Typical fluidic component characteristics can be illustrated by those for one of the most common fluidic amplifiers, the vented jet-interaction analog amplifier (Figure 4.2).

Graphically, the input characteristic of a single input port is a plot of

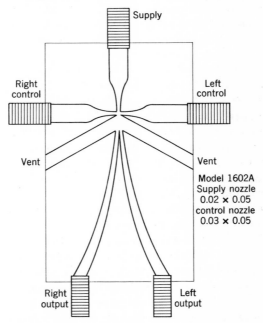

Figure 4.2 Description of a typical vented jet-interaction amplifier.

Graphical Characteristics   71

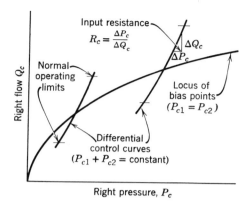

Figure 4.3 Typical static input characteristics of vented jet-interaction amplifier (one side only).

the control flow versus the pressure applied at the control port (Figure 4.3). In most vented amplifiers (which minimize internal feedback) the input characteristics are practically independent of output loading. Note that the locus of bias points is the curve generated when both control port pressures are always equal. The differential control curves result when one control port pressure is increased and the other decreased the same amount, keeping the *average* of the two always at a fixed bias pressure. These are the conditions under which the amplifier will normally be operated.

The transfer characteristic defines the gain of the amplifier and is represented by a family of curves with output load as the parameter (Figure 4.4). Note that it is normal for the pressure gain to decrease as the load impedance is reduced (opened from blocked conditions), and that beyond saturation there can be a *reversal* in slope.

The output characteristics are a plot of the output flow versus output pressure as the load is varied from near zero impedance (relatively large flow) to near infinite impedance (blocked output port) (Fig 4.5). Because the output characteristics (which define the output impedance) are also a function of the control signal, a complete description of the output characteristics is a *family* of curves of output flow versus output pressure with control pressure or control flow as the parameter.

Now note that the transfer characteristic and the output characteristics contain the same information; output behavior under load, in response to some input signal. Therefore, only one of these sets of curves is required. The output characteristics that are plotted in terms of flow and pressure are more convenient for analyzing the problems of cascading components; therefore, this form is preferred.

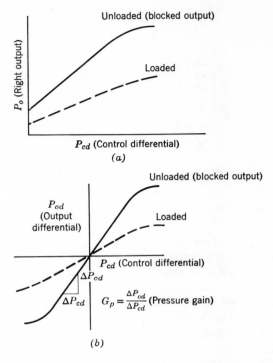

Figure 4.4 Typical transfer curves of vented jet-interaction amplifier; (a) single-ended; (b) differential.

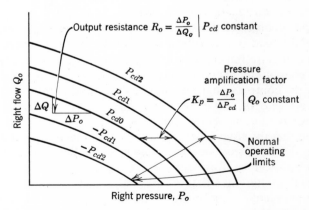

Figure 4.5 Typical static output characteristics of vented jet-interaction amplifier.

Graphical Characteristics 73

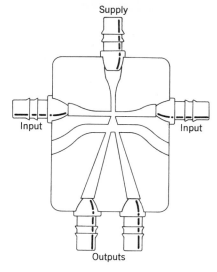

Figure 4.6 Description of a typical wall-attachment amplifier.

Digital Amplifiers

Typical digital fluidic component characteristics can be illustrated for the most common digital amplifier, the vented wall-attachment amplifier shown in Fig. 4.6.

Graphically the input characteristics of a single input port are a plot of the control flow versus the pressure applied at each control port, as shown typically in Figure 4.7. In this case (as with the proportional vented ampli-

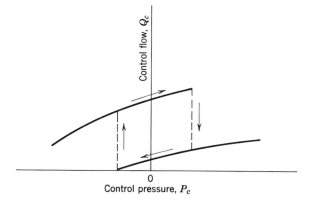

Figure 4.7 Input characteristics of wall-attachment amplifier.

74   Component Description

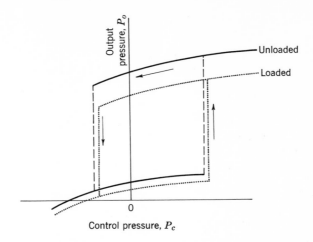

Figure 4.8 Switching characteristics of wall-attachment amplifier (one side only).

fier) the input characteristics are practically independent of output loading, but this may not be the case for other types. The most striking feature of the input characteristic is the abrupt discontinuity of the curve. This occurs at the point of switching, when the pressure has built to the point where it detaches the power stream from the adjacent wall allowing it to attach to the opposite wall, thereby increasing the impedance looking into that control port.

Note that the curve exhibits considerable hysteresis due to the latching effect of wall attachment; that is, the curve of increasing control pressure to the point of switching is different from the curve of decreasing control pressure to the point of reattaching. This curve is dependent on the pressure applied on the opposite control port.

The switching characteristic is shown in Figure 4.8. Assuming effective vents that prevent the feedback of output pressure into the interaction region, the switching characteristics should be shown as a family of curves with load as a parameter. Note that it is normal for the output pressure to decrease as load impedance is reduced; that is, opened from blocked conditions.

The output characteristics shown in Figure 4.9 are the output flow versus the output pressure as the load is varied from near zero impedance to near infinite impedance.

The output characteristics are a function of the control signal so a complete graphical description of the output characteristics of the digital amplifier can be illustrated by two curves, one representing the case when the power stream is deflected into the output leg being measured, and the

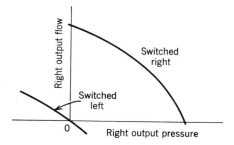

Figure 4.9 Output characteristics of wall-attachment amplifier.

other curve the case when the power stream is deflected away from the output leg being measured.

Note again that the switching characteristics of the digital amplifier and the output characteristics contain essentially the same information, but in a different form; that is, output behavior under load in response to an input signal. Therefore only one of these two sets of curves is required. The output characteristics are preferred because they are more convenient for analyzing the problems of cascading components.

Normalization of Characteristic Curves

The characteristics of fluidic devices are of course a function of supply pressure; therefore, to be complete, it is necessary to have a set of input and output characteristics for every allowable supply pressure. This can be done by providing individual input and output curves for a number of supply pressures and interpolating when necessary. But it is more

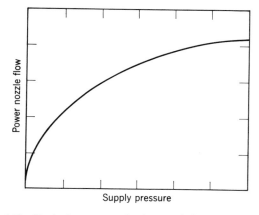

Figure 4.10 Typical power nozzle characteristic of fluidic amplifier.

convenient to do it by providing a single set of input and output characteristics *normalized* with respect to supply pressure and supply flow. In this case it is necessary to provide another characteristic curve defining how supply flow varies with supply pressure (essentially the power nozzle characteristic shown in Figure 4.10).

In summary, *the total static and large-signal characteristics of a fluidic device under normal operating conditions can be described by only three sets of curves; input characteristics including bias and differential curves, output characteristics with input signal as a parameter, and power nozzle pressure-flow characteristics.*

4.2 EQUIVALENT ELECTRIC CIRCUITS

Analog Amplifiers

As an equivalent electric circuit, the analog fluidic amplifier can be represented as shown in Figure 4.11. The input characteristics are described in terms of simple impedances between the control ports or between a control port and return. The transfer characteristics are represented by a pressure generator and a network whose output pressure is a function of the net pressure appearing at the control nozzle. The output characteristics are represented as simple series and shunt impedances that are directly coupled to the load impedance.

The elements of the equivalent electric circuits can all be calculated from the graphical characteristics, circuit dimensions, and conditions at the bias (quiescent, no-signal) operating point.

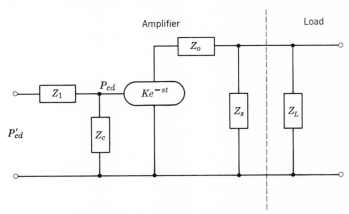

Figure 4.11 Typical electric equivalent circuit for analog amplifier.

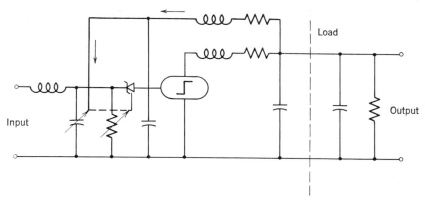

Figure 4.12 Typical electric equivalent circuit for digital wall-attachment amplifier.

Digital Amplifiers

As an equivalent electric circuit the digital fluidic amplifier can be represented as shown in Figure 4.12. The input characteristics are described in terms of nonlinear impedances between a control port and return, which are controlled by output conditions.

The switching characteristics are illustrated by a pressure generator with infinite gain, and an output-controlled reference diode at the input to the pressure generator. The output characteristics are represented as simple linear series and shunt impedances that are directly coupled to the load impedance.

The elements of the electrical equivalent circuits can all be calculated from the graphical characteristics, the circuit dimensions, and the conditions at the bias operating point.

In summary, *the small-signal and dynamic characteristics of a fluidic device can be represented by an equivalent electric circuit containing various linear and nonlinear impedances and a simple generator.*

4.3 IMPORTANT PHYSICAL QUANTITIES

In the calculation of the dynamic behavior of fluidic components, it is necessary to take into account certain physical dimensions of the circuit and certain qualities of the operating fluid.

Line Lengths

The lengths of lines carrying pressure waves are important in the calculation of equivalent inductance (or inertance) and resistance. This includes the lengths of interconnecting lines, coupling hardware, and passages inside the fluidic component.

Jet Path Lengths

In passages and chamber areas where pressure waves will not propagate and the fluidic signal is transmitted as a flow variation (as in a free submerged jet), it is necessary to know the lengths of paths. From this it is possible to calculate the transit time of a signal through a fluidic component.

Effective Areas

In the calculation of equivalent inductance (inertance) and resistance, it is necessary to consider the effective area of the length of passage. For a fluidic component this may be presented for each different area or as an effective area based on the formula [from Ref.(8)],

$$A_{\text{effective}} = \frac{(A_{\text{in}} - A_{\text{out}})}{\ln (A_{\text{in}})/(A_{\text{out}})}$$

where A_1 and A_2 are the areas of inlet and outlet sections. The effective areas of all control and outlet passages, coupling devices, and interconnecting lines should be given.

Effective Volumes

In the calculation of equivalent volume capacitances in the circuit containing the fluidic component, it is necessary to know the effective volume of fluid trapped at every level of static pressure; that is, the total volume of fluid under compression. Effective volumes of all lines, connectors, and internal passages must be given.

Power Jet Velocity

In the calculation of transit time, it is necessary to know the average velocity of the power jet in the interaction chamber. Either this can be measured with a Pitot tube inserted up through a receiver or it can be calculated from the velocity of the power jet as it exits from the nozzle and as it enters the receiver.

Qualities of the Operating Fluid

In the calculation of equivalent inductance and equivalent line resistances, and in the definition of the conditions under which static characteristics are presented, it is important to know the qualities of the operating fluid. Unless the operating conditions are other than standard, ambient temperature, return and vent pressures, and type of fluid should be provided.

4.4 PERFORMANCE PARAMETERS

Performance parameters can be defined for two basic purposes: first, to *describe the behavior* of a device under static or dynamic conditions (like pressure gain) and second, to provide the data necessary to *calculate* the behavior from basic information (like output impedance). The performance parameters most pertinent to fluidic control systems are defined in the following paragraphs.

Output Resistance R_o

Output resistance is defined as the ratio of a change in output pressure to a change in output flow for a fixed control signal, that is,

$$R_o = \frac{\Delta P_o}{\Delta Q_o}\bigg|_{P_{cd}\,\text{constant}}$$

With reference to the static output characteristics of a typical amplifier (as shown in Figure 4.5), the output resistance is simply the slope of one of the family of curves. Thus the output characteristic curves define the output resistance under *all* static conditions, but because the characteristic curves are not linear, the actual output resistance is quite variable. Therefore in determining the appropriate numerical value, the resistance must be calculated at the point at which the amplifier is operated when connected in a circuit.

Pressure Gain G_p

Pressure gain is defined as the ratio of the change in output pressure to the change in control pressure, when the fluidic amplifier is operating in a particular circuit (with a particular load). That is (for a differential amplifier),

$$G_p = \frac{\Delta P_{od}}{\Delta P_{cd}}$$

80   Component Description

The transfer curve for a typical differential amplifier is shown in Figure 4.13. By the above definition the pressure gain is the slope of the transfer curve. Since the curve is not linear, the point at which the amplifier operates in a circuit must be specified in calculating a numerical value for pressure gain.

In the case of the digital fluidic amplifier, the definition of pressure gain is the ratio of the change in output pressure to the change in control pressure required *for switching to occur*. That is (see Figure 4.13),

$$G'_p = \frac{P_{od(sw)}}{P_{cd(sw)}}$$

It should be recognized that, according to this definition, the gain of the digital amplifier can be infinite if it has negligible hysteresis.

Pressure Amplification Factor K_p

The pressure amplification factor for amplifiers is defined as the ratio of the change in output pressure to the change in control pressure *when the output flow is constant.* That is,

$$K_p = \frac{\Delta P_o}{\Delta P_{cd}}\bigg|_{Q_o \text{ constant}}$$

In effect, this is the maximum pressure gain that an amplifier could deliver if there were no loading effects (zero amplifier output resistance). With reference to the output characteristic curves for a typical amplifier (shown in Figure 4.5) one can see that the amplification factor is a func-

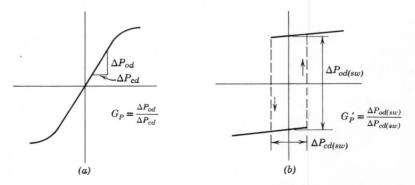

Figure 4.13   Definition of parameters from transfer characteristics; (*a*) analog; (*b*) digital.

tion of the horizontal distance between the output impedance curves. And since the curves are neither linear nor evenly spaced, it is evident that the pressure amplification factor is quite variable. Therefore, in determining the appropriate numerical value for K_p, calculations must be made in the vicinity of the point at which the amplifier operates in a circuit.

Flow Gain G_f

Flow gain is defined as the ratio of the change in output flow to the change in control flow, when the fluidic amplifier is operating in a particular circuit (with a particular load). That is (for a differential amplifier),

$$G_f = \frac{\Delta Q_{od}}{\Delta Q_{cd}}$$

By this definition the flow gain is the slope of the transfer curve plotted in units of flow. Since the curve is not linear (similar to the pressure transfer curve of Figure 4.13), the point at which the amplifier operates in a circuit must be specified in calculating a numerical value for flow gain.

Sensitivity Factor K

A sensitivity factor K can be defined for any fluidic sensor in the same way that the pressure amplification factor K_p is defined for an amplifier. The only difference is that the input variable is the quantity that the sensor measures instead of the control differential pressure. That is,

$$K = \frac{\Delta P_o}{\Delta \text{ sensed variable}} \bigg|_{Q_o \text{constant}}$$

Since most output characteristic curves for fluidic sensors are nonlinear, the numerical value for the sensitivity factor must be calculated at the point at which the sensor operates when connected in a circuit.

Input Resistance R_c

Input resistance is defined as the ratio of the change in control pressure to the change in control flow when the bias pressure is held constant. That is,

$$R_c = \frac{\Delta P_c}{\Delta Q_c} \bigg|_{(P_{c1} + P_{c2}) \text{constant}}$$

With reference to the typical differential amplifier static input characteristics (Figure 4.3) the input resistance is simply the slope of the appropriate curve. And since the curves are nonlinear, the numerical value of the input resistance must be calculated at the point at which the amplifier operates when connected in a circuit.

Equivalent Capacitance C

Because of the compressibility of the operating fluid, there is an equivalent capacitor formed by every element of volume under pressure in the fluidic circuit. As a result, the change of pressure at every point is delayed until there is sufficient flow to satisfy the conditions of compressibility at the new pressure level. The effect is analogous to an electrical shunt capacitor and can be treated as such in equivalent circuit analysis (see Chapter 8).

The equivalent capacitance of a fluidic device is defined as the ratio of the trapped volume V to the absolute static pressure P_a. That is,

$$C = \frac{V}{P_a} \quad \text{(for isothermal flow)}$$

Since the passages are seldom uniform and the pressure is not the same in every section, each must be calculated as a separate element; then they must be added together to arrive at a total circuit capacitance. The pressure used in the calculation of the equivalent capacitance must of course correspond with the point at which the device operates in a circuit.

In digital wall-attachment devices the major portion of the effective capacitance is due to "charging" the attachment "bubble."

Equivalent Inductance L

Because of the inertance of the operating fluid, there is an equivalent inductor formed by every element of mass in the fluid circuit. As a result, the change in flow at every point is delayed until sufficient forces can build up and accelerate the flow to the new level. The effect is analogous to an electrical series inductor and can be treated as such in an equivalent circuit analysis (see Chapter 8).

The equivalent inductance of a fluidic device is defined as the ratio of the product of mass density and length to the effective cross-sectional area. That is,

$$L = \frac{\rho l}{A_{\text{eff}}}$$

Since the area of the passages in fluidic circuits is seldom uniform and the density is not the same in every section, each must be calculated as a separate element; then they must be added together to arrive at a total circuit inductance. The pressure used to calculate mass density must of course correspond with the point at which the device operates in a circuit.

Time Delay t_d

The total time delay is made up of a transit time t'_d and a propagation time t''_d. The transit time delay in a fluidic device is defined as the time required to transport a deflected element of fluid in the power stream from the power nozzle to a point just inside the receiver where pressure waves can propagate. It can be calculated from the ratio of the length of the interaction chamber (where the power jet is essentially free) to the average velocity of the power stream. That is,

$$t'_d = \frac{\text{length of interaction chamber}}{\text{average velocity of fluid flow}}$$

The total time delay also includes the delay in the control and outlet passages due to the finite velocity of sound. In those cases, the delay is calculated from

$$t''_d = \frac{\text{length of aperture}}{\text{velocity of sound} + \text{velocity of steady fluid flow}}$$

and the *total time delay* is

$$t_d = t'_d + t''_d$$

Pressure Recovery Factor R_p

Because of losses in fluidic amplifiers, it is not practical to recover 100% of the total power supplied to it. In many cases the recovery is an important factor in the selection of an amplifier for a specific application. Therefore, we define the pressure recovery factor as the ratio of the *maximum* output pressure to the supply pressure. That is,

$$R_p = \frac{(P_o)_{\max}}{P_s}$$

Note that in the normal differential amplifier the maximum output pressure is recovered when the power stream is in a *deflected* condition. Note also that similar recovery factors can be defined for flow and power.

Signal-to-Noise Ratio S/N

Signal-to-noise ratio in fluidic circuits has the same meaning that it has in electronic circuits. Expressed in decibels, it is the ratio of the *maximum* signal capability to the noise at one point in the circuit. That is,

$$S/N = 20 \log_{10} \frac{\text{maximum useable signal}}{\text{noise}}$$

In fluidic amplifiers, the major source of noise is from turbulence and instability of the power stream; therefore, a reasonable measure of the noise can be made at the operating bias point.

5
Test Methods and Instrumentation

5.1 STATIC

The circuit for measuring the static characteristics of a differential fluidic amplifier is shown in Figure 5.1. Note that flowmeters are required for one output side only, providing that the second output port is connected to a dummy load with the same impedance as the flowmeter.

The circuit contains, in addition to the amplifier under test, a regulated supply of air, manually controlled valves in the lines leading to the control ports, and identical manually controlled valves in the lines from the output ports. The flowmeter in the output circuit should be chosen for low pressure drop. The pressure meters at the supply, control, and output ports can be simple U-tube manometers containing a fluid (mercury or water) suitable for the pressures being measured. Expansion tanks are used at each port to measure the approximate total pressure. Data can be recorded directly onto properly scaled graphs labeled output flow versus output pressure and input flow versus input pressure.

Analog Amplifiers

The analog amplifier is tested as follows: a nominal supply pressure (one for which the amplifier was designed) is applied, and the pressures at the control ports are adjusted to 10% of the supply pressure. (This is the control bias pressure where most jet-interaction amplifiers perform best, but other bias pressures may be appropriate for a particular case.)

The first run (with wide-open load valve) is made by varying the control differential pressure from zero to +10% of supply pressure and from zero

86 Test Methods and Instrumentation

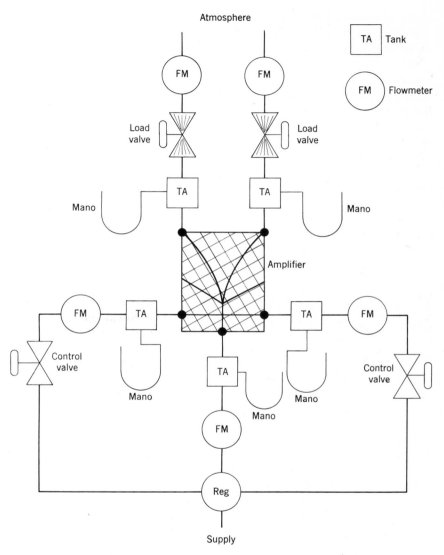

Figure 5.1 Circuit for testing differential fluidic amplifier.

to −10% of supply pressure. Note that the bias (the average of the two control pressures) must always remain at 10% of supply pressure. For example, when the right control pressure is raised to 11%, the left control pressure is lowered to 9%, thereby generating +2% control *differential* while maintaining the bias at 10%. At each point of control differential, the output flow and pressure define a point on the output graph and the input flow and pressure define a point on the input graph.

When the first run is complete, the load circuit valves are closed an equal amount, giving roughly $\frac{1}{4}$ of the supply pressure at the output ports with zero control differential. The second run, varying the control differential pressure from zero to $+10\%$ of supply and from zero to -10% of supply, yields a new set of points on the output graph. The input points should be checked against the results of the first run to determine if it is necessary to record them.

Additional runs are made in an identical manner: the load valves are closed in increments until the final run with blocked output ports.

Finally, points of equal control differential pressure are connected with smooth curves. A typical set of output characteristic curves for a jet interaction amplifier is shown in Figure 5.2. Note that, although they represent a differential amplifier, they are actually measured on one side only, if the amplifier is reasonably well balanced.

A second and third set of output and input characteristics should be taken with the same supply pressure but with control bias pressures of 5% of supply and 20% of supply, respectively. A fourth and fifth set, taken at supply pressures above and below the nominal, and with control bias at 10% of supply, will complete the characteristics necessary for matching operating points and analyzing the response to large signals.

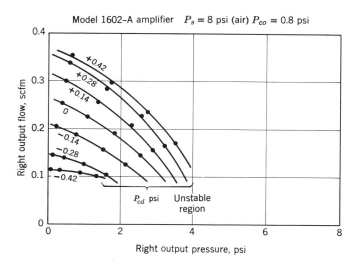

Figure 5.2 Static output characteristics of a jet-interaction amplifier.

Digital Amplifiers

The static characteristics of a differentially connected digital amplifier are generated in a manner somewhat different from that used for the analog amplifier. This is because (1) we are mainly interested in the characteristics around the switching points and (2) it is often required to operate in the region where the output circuit is returned to a reference pressure below atmospheric.

In the case of the digital amplifier, the testing procedure is as follows: a nominal supply pressure is applied to the test circuit shown in Figure 5.1, the loads are returned to a regulated pressure of approximately -5 psig, and the pressures at the control ports are adjusted to a recommended bias level (nominally 10% of supply pressure).

The first run, with wide-open load restrictors, is made by varying the control differential from $+10\%$ of supply pressure to -10% of supply pressure, then from -10% *back to* $+10\%$. Note that it is important to advance in one sense to a switching point, then in the opposite sense to the reverse switching point in order to measure the hysteresis accurately.

At each point of control differential, the output pressure and flow define a point on the output characteristic, the input pressure and flow define a point on the input characteristic, and the power nozzle pressure and flow define a point on the power nozzle characteristic.

Additional runs are made similar to the analog case, readjusting the load impedance for each run until the loads are totally blocked. The data recorded are sufficient to generate static input, output, and power nozzle characteristics.

The static characteristics of other types of analog and digital fluidic amplifiers are measured in a similar manner. It should be emphasized that, when many characteristics are to be recorded, it is most convenient to automate the process, using electronic transducers and a graphical XY plotter.

Vortex Rate Sensor

The circuit for measuring the static output characteristics of a push-pull vortex rate sensor is shown in Figure 5.3. Note that a flowmeter is required in only one of the output lines, providing that the adjacent line contains an equivalent dummy impedance.

The circuit contains, in addition to the rate sensor under test, a regulated supply of air, a means for rotating the sensor at a known rate increment, manually controlled valves in the output, a flowmeter of low pressure drop, and U-tube manometers containing appropriate fluids (typically water).

Static Test Methods 89

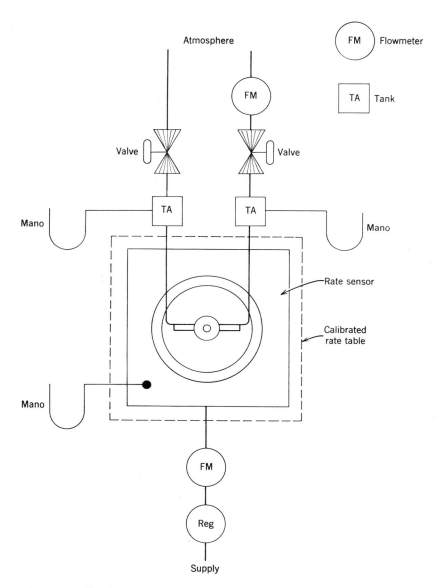

Figure 5.3 Circuit for measuring static output characteristics of vortex rate sensor.

The rate sensor is tested as follows. A nominal supply pressure is applied, and the load valve is opened wide. The rate sensor is rotated at a known rate, and the output conditions are recorded as a point on an appropriately scaled graph of output flows versus output pressure. The rate of turn is changed by a known increment, and the resulting output circuit conditions are recorded as a point on the output graph.

When a predetermined number of values of rate of turn have been tested, the load circuit valves are closed by equal increments, the range of rate increments is repeated, and the resulting output pressures are recorded. Additional runs are made with load valves set at increments giving uniformly spaced points on the output graph until the loads are completely blocked. Finally, points of equal rates of turn are connected with smooth curves. The result is a set of output characteristics, shown in Figure 5.4, which are typical of many vortex rate sensors.

The static characteristics of other types of fluidic sensors are measured in a similar manner.

Impedances

The circuit for measuring the static characteristics of any fluidic passive element is shown in Figure 5.5. The circuit contains, in addition to the element under test, a regulated supply pressure, a manually controlled valve, a U-tube manometer connected across the resistor, and a low-pressure-drop flowmeter on the exhaust side of the circuit.

A resistor is tested as follows. The supply pressure is set at a value slightly above the maximum that the resistor will be exposed to while in use. The manual valve is opened wide, resulting in a particular combination of pressure drop and flow that is recorded as a point on an appropriately scaled graph of flow versus pressure drop. The manual valve is

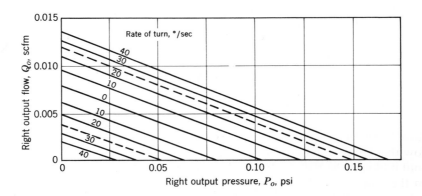

Figure 5.4 Static output characteristics of a vortex rate sensor.

Static Test Methods 91

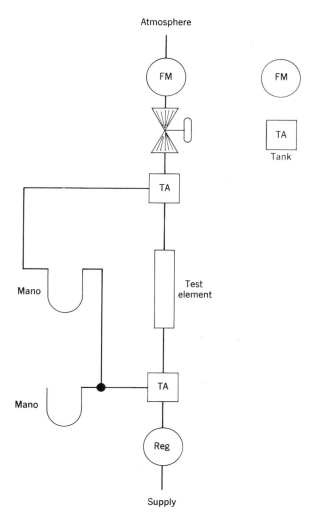

Figure 5.5 Circuit for measuring static characteristics of a passive element.

closed a given increment, causing a new combination of pressure drop and flow that is recorded as a new point. The valve is closed by increments, until it blocks the flow to the element, and the resulting points are plotted on the graph. Finally, the points are connected to form a smooth curve representing the static characteristics of the element. The result is a characteristic typically as shown in Figure 5.6.

92   Test Methods and Instrumentation

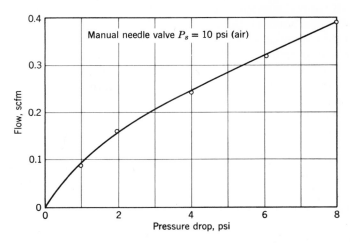

Figure 5.6   Static characteristics of a fluidic resistor.

5.2 DYNAMIC

Analog Amplifiers

The circuit for measuring the dynamic characteristics of a push-pull (differential) fluidic amplifier is shown in Figure 5.7. Note that the use of pressure instrumentation is illustrated. Flow transducers could be used if preferred.

The test circuit contains, in addition to the amplifier under test, a

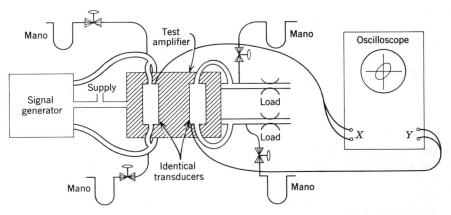

Figure 5.7   Circuit for measuring dynamic characteristics of a differential amplifier.

source of supply, a pneumatic sinusoidal signal generator with separately adjustable bias and amplitude levels (to be described in detail later in this chapter), manometers containing appropriate fluids connected at the control and output ports, equal loads connected at the output ports, identical minimum-volume differential pressure transducers connected between the control ports and between the output ports, an electrical power supply (typically two 3-volt dry-cell batteries), and a high-sensitivity XY plotting oscilloscope with identical X and Y amplifiers. The X axis is connected to the input transducer, and the Y axis is connected to the output transducer.

The amplifier is adjusted for test as follows. Supply pressure (typically 8 psig) is applied to the amplifier circuit and the pneumatic sinusoidal signal generator. The signal generator is set to run at a speed producing 5 to 10 Hz, while the control circuit bias pressure is adjusted to 10% of the supply pressure (or any other appropriate bias level). The signal generator drive motor is then shut off, and by advancing the generator rotor disc slowly by hand, the amplitude of the sinusoidal differential pressure signal is measured. The amplitude is then adjusted (typically 10% of supply pressure, peak to peak) to the proper level. If an adjustment is made, it is necessary to repeat the bias adjustment again, and then the amplitude adjustment, until both requirements are satisfied. The bias and the amplitude pressures are noted, and the control circuit manometers are cut off. The load bias pressures are noted, and the load circuit manometers are cut off.

With the signal generator running at low speed (typically 1 to 2 Hz) and the transducer power supply connected, the oscilloscope sensitivities are adjusted (in the XY mode) for good readability (as illustrated in Figure 5.8a). The figure on the oscilloscope is obviously a plot of the output pressure (Y axis) versus the input pressure (X axis), and if the oscilloscope amplifiers have been set for equal sensitivity, the slope of the figure is the pressure gain of the amplifier circuit.

At higher frequencies, when there is a significant amount of phase shift between the output signal (Y axis) and the input signal (X axis), the figure appearing on the oscilloscope is not a single line, but rather a closed figure. If the amplifier (or circuit element) under test is ideally linear, the figure is a straight line for zero phase difference, an ellipse with its major axis greater than the slope of the straight line for phase differences between 0 and 90°, and an ellipse with its major axis exactly vertical for a phase difference of 90°. For phase differences greater than 90°, the major axis of the ellipse rotates into the second, third, and fourth quadrants. These situations are illustrated in Figure 5.8 for phase differences of less than 180°.

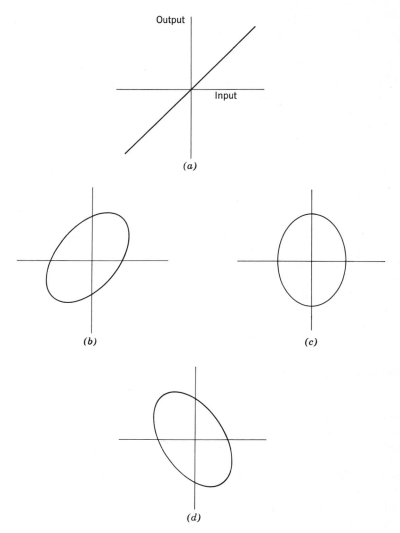

Figure 5.8 Typical oscilloscope displays of sinusoidal response (XY mode); (a) in-phase; (b) 0 to 90° phase lag; (c) 90° phase lag; (d) 90 to 180° phase lag.

Both gain and phase difference can be read from the oscilloscope display. Gain is the ratio of the maximum excursion in the Y direction to the maximum excursion in the X direction, as illustrated in Figure 5.9. The phase difference is calculated from the ratio of the width of the ellipse in the Y direction to the maximum excursion in the Y direction, that is,

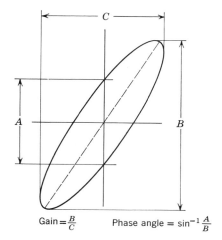

Figure 5.9 Measuring gain and phase angle from oscilloscope display.

$$\text{sine of phase angle difference} = \frac{\text{width of ellipse}}{\text{maximum } Y \text{ deflection}}$$

or

$$\theta = \sin^{-1}\frac{A}{B}$$

Frequency Response Testing

The amplifier circuit is now ready for frequency response testing. The first point is measured with the signal generator set at a very low frequency (less than 1 Hz). The gain and phase shift are calculated from the figure on the oscilloscope. The frequency (which can also be measured on the oscilloscope by temporarily switching to the time base mode of display), the gain, and the phase shift are recorded. Other points are measured at convenient increments of frequency (typically 1, 2, 5, 7, 10, 20, etc.) up to frequency where the output is so attenuated that the response cannot be separated from the noise appearing on the oscilloscope.

From these data, the gain is calculated in decibels = $20 \log_{10} P_{out}/P_{in}$. The results are plotted as a familiar Bode diagram (Figure 5.10) which shows how the magnitude of the gain and the phase shift of the amplifier circuit vary with frequency of the incoming sinusoidal pressure signal.

Measuring Total Time Delay

For measuring the time delay (transport time) of a fluidic amplifier, it is convenient to substitute a square-wave signal generator for the sinusoidal

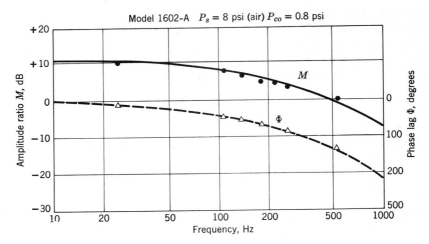

Figure 5.10 Frequency response of a proportional amplifier.

signal generator used in the frequency response tests (Figure 5.7). The transient response is then displayed on the oscilloscope used in its two-trace mode (both input and output pressures displayed on a common time base). A typical response is shown in Figure 5.11. The time delay (transport time) is then measured as the time difference between the *initial* rise of the square-wave input signal and the *initial* rise of the output

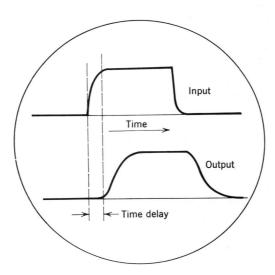

Figure 5.11 Typical oscilloscope display of step response (common time base).

pressure. The measurement is made more convenient by photographing the oscilloscope display and reading the resulting print.

The time delay appears in the frequency response as a phase difference directly proportional to frequency. The phase contributed by the time delay can be calculated from

$$\theta = 360 \times \text{time delay} \times \text{frequency, degrees}$$

Digital Devices

For defining the switching time of digital fluidic devices we are usually concerned only with conditions around the switching points. In this case the square-wave signal generator is connected to the test circuit and adjusted to provide just enough signal excursion to ensure positive switching. The response is then displayed on the oscilloscope used on the two-trace mode (both input and output pressures displayed on a common time base). The response is similar to that shown in Figure 5.12. The switching time is measured as the time delay from the *initial* rise of the *input* signal until the *output* reaches 90% of its final value.

Note that in the response there is a dead time until the attachment bubble (or equivalent capacitance) is charged, then the output rises to the new level in an exponential fashion. Note also that in many cases the turn-on time will be considerably different from the turn-off time.

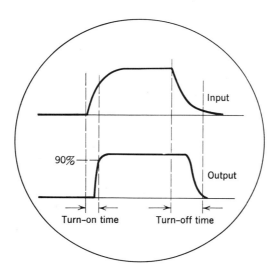

Figure 5.12 Typical oscilloscope display of switching time (common time base).

Sensors

The circuit for measuring the dynamic characteristics of a typical fluidic sensor (rate sensor) is shown in Figure 5.13. The circuit contains in addition to the sensor under test, a source of supply pressure, a means for introducing a dynamic change in the sensed variable (such as a sinusoidally driven rate table), a transducer to measure the sensed variable independently (such as a high-performance electromechanical rate gyro), a mini-

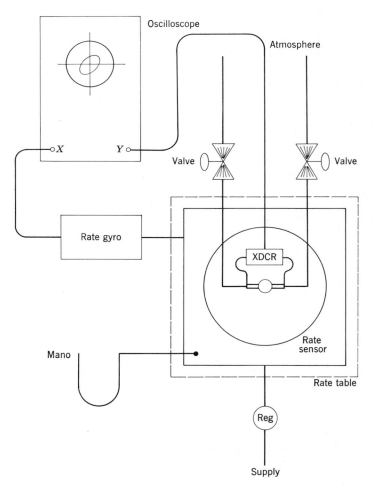

Figure 5.13 Circuit for measuring dynamic characteristics of a vortex rate sensor.

mum volume differential pressure transducer, a source of power for the transducers, an appropriate load on the rate sensor outputs, and an XY oscilloscope.

The input transducer is connected to the X axis of the oscilloscope, and the output transducer is connected to the Y axis of the oscilloscope. With the oscilloscope in the XY mode the input is varied sinusoidally at a relatively small peak-to-peak amplitude at various frequencies of interest.

The amplitude and phase response of the sensor can then be measured directly from the oscilloscope display as described in the section on "Dynamic Testing of Analog Amplifiers." The results are plotted as a familiar Bode diagram showing the variation of amplitude and phase with frequency.

The time delay (transport time) of the sensor is measured from the response to a step input. The results are displayed on the oscilloscope in the two-trace mode with both input and output signals plotted on a common time base. The time delay is measured as the difference between the initial rise of the input signal and the initial rise of the output signal. The measurement is more convenient if the oscilloscope display is photographed and the reading taken from the resulting print.

Impedances

The circuit for measuring the dynamic characteristics of any passive fluidic element is shown in Figure 5.14. In addition to the element being tested, the circuit requires a source of fluid supply, a sinusoidal differential pressure signal generator, a minimum-volume differential pressure transducer, a source of power for the transducer (3-volt battery), a miniature dynamic flow probe (hot-wire anemometer), and an XY plotting oscilloscope.

In running a frequency response test of an impedance element, it may be desirable to have a biased differential pressure signal, simulating the conditions encountered in most fluidic amplifier circuits. The signal generator can be unbalanced to provide this bias flow.

To measure the dynamic characteristics, the flow transducer is connected to the Y axis of the oscilloscope and the pressure transducer is connected to the X axis. The signal generator is adjusted for a peak-to-peak amplitude representing typical conditions (say, 1 psi). The signal generator is then set to run at a low frequency (below 1 Hz). The figure displayed on the oscilloscope in the XY mode is obviously a plot of the pressure-flow static characteristics. (Typically, an orifice would display a square-law relationship; a capillary would show a linear relationship.)

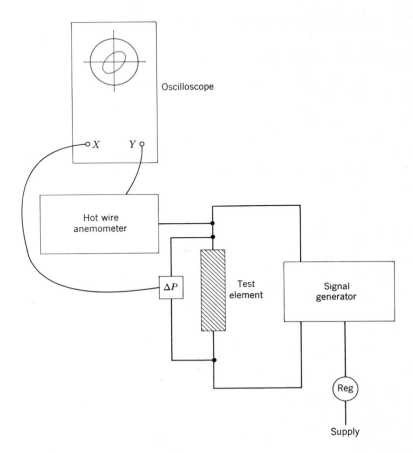

Figure 5.14 Circuit for measuring dynamic characteristics of a passive element.

Thus, the impedance of the passive element would be a function of the slope of the curve around the bias point. With properly calibrated transducers, the impedance at that frequency can be calculated directly from the oscilloscope display.

The frequency can be increased in convenient increments, and the magnitude and phase angle of the impedance can be measured from the oscilloscope display as described earlier. The results are plotted in the form of a Bode diagram, showing how the magnitude of the impedance and the phase angle of the impedance vary with frequency of the applied pressure.

5.3 SPECIAL TEST EQUIPMENT

Pneumatic Pressure Signal Generator

The pneumatic signal generator required to implement the foregoing dynamic tests must have the following specifications:

1. Deliver a differential pressure signal.
2. Include an independently adjustable bias pressure in both output lines.
3. Include an adjustment for peak-to-peak amplitude of the output signal.
4. Have a continuously adjustable frequency from 1 to 300 Hz.
5. Generate either a sine wave or a square wave ($\pm 5\%$ distortion) with a minimum of changeover effort.
6. Maintain a relatively constant ($\pm 10\%$) output over the specified frequency range.
7. Require standard pneumatic and electrical power inputs.

Signal generators meeting these specifications generally use a motor-driven wobble plate with two impinging back-pressure nozzles.

Pressure Transducers

The pressure transducers necessary to implement the test circuits such as those shown in Figures 5.7 and 5.13 have one major physical requirement. That is, they must be capable of measuring pressures in a physical circuit without introducing a large volume capacitance. Therefore, they must have a minimum-diameter flush-mounted diaphragm and be connected into the test circuit by means of a minimum-volume adapter similar to the one illustrated in Figure 5.15 using the shortest possible lines. It should be noted that in a differential pressure transducer, the dynamics of the two sides are not identical and can never be made so. Therefore, they should be connected into the test circuit in the same phase (i.e., high pressure on the same sides at the same time) so that the effect will be partially cancelled.

Flow Transducers

For dynamic testing, the only practical flow sensor known is the hot wire. However, there are many levels of sophistication in the instrumentation circuits necessary to provide an indication of the instantaneous

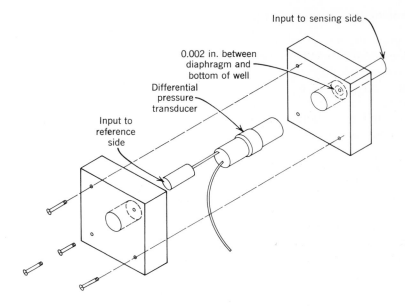

Figure 5.15 Pressure transducer with minimum-volume adapter.

flow. Again, since the dynamic behavior of any fluidic circuit is so highly dependent on the associated lines and volumes, the hot-wire sensor should be arranged so that is introduces a minimum of additional volume and restriction into the test circuit. The sensors should also be properly phased to minimize any resulting nonlinear effects. That is, they should be so arranged that high flow occurs at the same instant at each sensor.

XY Oscilloscope

For measuring the frequency response characteristics of fluidic devices, it is most convenient to use an oscilloscope with the following characteristics:

1. Two beams synchronized to common time base.
2. Calibrated time base.
3. Identical amplifiers on both channels.
4. High sensitivity (2 mV/cm).
5. Convertible for *XY* plotting.

Automated Equipment

Through the use of more sophisticated automatic equipment the testing process can be made more convenient, more accurate, and faster. With pressure and flow transducers connected to an XY plotter, graphical characteristics can be plotted directly. In dynamic testing, the use of a transfer-function analyzer provides for direct readout of magnitude ratios and phase angles and for direct continuous plotting of the Bode diagram.

6

Graphical Characteristics of Typical Fluidic Devices

Graphical characteristics for general classes of active fluidic devices were defined and discussed in Chapter 4. In this chapter actual graphical characteristics of typical active and passive fluidic devices are illustrated. They will show what is to be expected in the "real world."

6.1 TURBULENT (NONLINEAR) RESTRICTORS

Turbulent restrictors are most common in fluidic circuits. Various types operate with various degrees of turbulence; therefore, their characteristics differ. Shown in Figure 6.1 are three typical examples of the characteristics of turbulent restrictors: a sharp-edged orifice, a standard needle valve, and a short length of plastic tubing. Note that the characteristics of the most turbulent, the sharp-edged orifice, closely approach a square law. That is,

$$Q^2 = K(\Delta P)$$

Since incremental resistance is defined from the slope of these curves it is important to note that the numerical value of resistance is not constant (the resistance is nonlinear). It must be calculated at the point where it is to be operated in a circuit.

6.2 LAMINAR (LINEAR) RESTRICTORS

Laminar restrictors are sometimes needed in a fluidic circuit to avoid the effects of nonlinear resistance. Various types are used, all of them based

Laminar (Linear) Restrictions   105

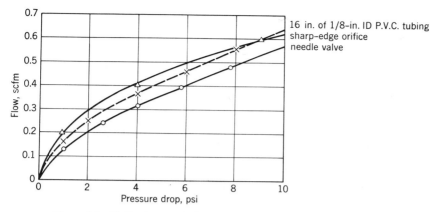

Figure 6.1 Static characteristics of typical turbulent-flow restrictors.

on laminar flow through very small passages. Shown in Figure 6.2 are typical characteristics of several available types: Fotoceram;* extruded ceramic rod; and bundled stainless steel capillary tubes. Note that over a considerable range of pressure, the characteristic of the bundled capillary tubes is a straight line. That is,

$$Q = K(\Delta P).$$

Since resistance is defined from the slope of the characteristic curve, the

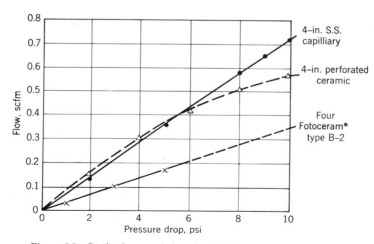

Figure 6.2 Static characteristics of typical laminar-flow restrictors.

*Trademark of Corning Glass Works.

resistance is constant over this range; therefore, the specific point of operation need not be known (so long as it is within the linear range).

6.3 ANALOG AMPLIFIERS

Vented Jet-Interaction Amplifier Output Characteristics

The output characteristics of a typical vented jet-interaction differential amplifier are shown in Figure 6.3. Note that the amplifier is not stable with blocked loads.

Vented Jet-Interaction Amplifier Input Characteristics

The input characteristics of a typical vented jet-interaction amplifier are shown in Figure 6.4.

Vented Jet-Interaction Amplifier Transfer Characteristics

The transfer characteristics of a typical jet-interaction amplifier are shown in Figure 6.5. Note that these apply to only one circuit connection; if the circuit is changed, the transfer characteristics change.

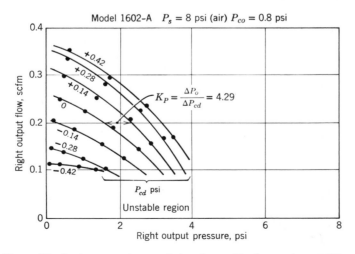

Figure 6.3 Static output characteristics of vented jet-interaction amplifier.

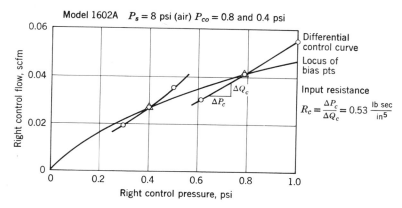

Figure 6.4 Static input characteristics of vented jet-interaction amplifier.

Closed Jet-Interaction Amplifier Output Characteristics

The output characteristics of a typical closed jet-interaction amplifier are shown in Figure 6.6. Note that by comparison with Figure 6.3 the use of vents avoids the complete loss of gain (distance between curves) at low output flows.

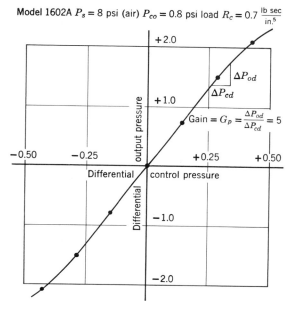

Figure 6.5 Static transfer characteristics of vented jet-interaction amplifier.

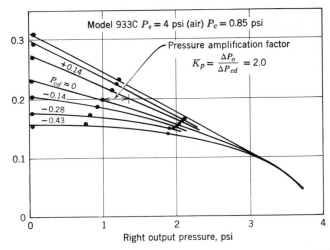

Figure 6.6 Static output characteristics of closed jet-interaction amplifier.

Note that in the closed amplifier, the effects of loading are reflected back to the input ports. Therefore, the input characteristics will be a *family* of curves similar to Figure 6.4 with load resistance as the parameter.

Vented Elbow Amplifier Output Characteristics

The output characteristics of a typical vented elbow amplifier are shown in Figure 6.7. Note that this type of amplifier is characterized by very low output resistance ($\Delta P_o/\Delta Q_o$).

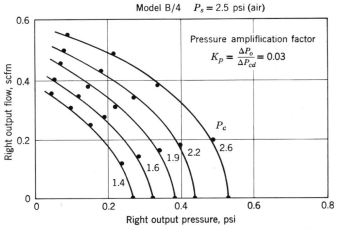

Figure 6.7 Static output characteristics of vented elbow amplifier.

Digital Amplifiers 109

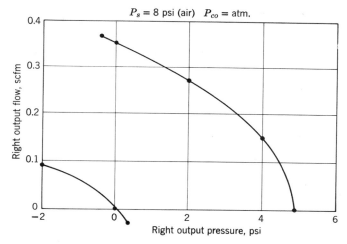

Figure 6.8 Static output characteristics of vented wall-attachment flip-flop.

6.4 DIGITAL AMPLIFIERS

Vented Wall-Attachment Flip-Flop Output Characteristics

The output characteristics of a typical vented wall-attachment flip-flop are shown in Figure 6.8. Note that there are only two curves, one defining the conditions when the flow is switched toward the leg being measured and the other defining the conditions when the flow is switched *away* from the leg being measured. Note also the branch of the curve in the negative pressure region, generated by applying a vacuum at the output port.

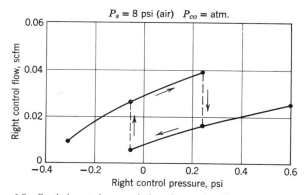

Figure 6.9 Static input characteristics of vented wall-attachment flip-flop.

Vented Wall-Attachment Flip-Flop Input Characteristics

The input characteristics of a typical vented wall-attachment flip-flop are shown in Figure 6.9. Note that the curve extends into the negative region when the power stream is attached to the side being measured. This is necessary to maintain the negative pressure in the attachment bubble.

Turbulence Amplifier Output Characteristics

The output characteristics of a typical turbulence amplifier are shown in Figure 6.10. Note that the pressures involved are quite low because the supply pressure must be kept low to produce a laminar jet. Note also that the pressure recovery is well over 50%, an outstanding feature of the turbulence amplifier.

Turbulence Amplifier Input Characteristics

The input characteristics of a typical turbulence amplifier are shown in Figure 6.11. Note that the input impedance is relatively high and there is no cross-coupling from either the output or from adjacent control nozzles. These features contribute to the turbulence amplifier's reputation for easy interconnection.

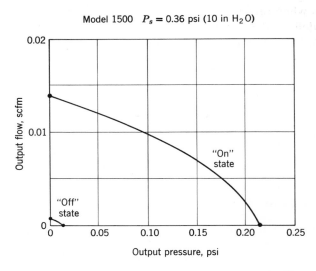

Figure 6.10 Static output characteristics of turbulence amplifier.

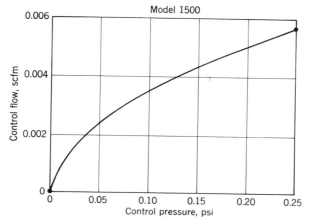

Figure 6.11 Input characteristics of turbulence amplifier.

6.5 SENSORS

As described in Chapter 4, the characteristics of sensors can be illustrated by a set of output curves with the sensed variable as a parameter.

Back-Pressure Nozzle Output Characteristics

The circuit for the distance-sensing back-pressure nozzle includes a network of fixed restrictors, so the output characteristics will reflect the characteristics of the restrictors selected. Figure 6.12 shows the output characteristics of a sensing nozzle with a circuit containing a typical set. Note that with a high resistance load, the output pressure is a nonlinear function of the distance from nozzle to object.

Interruptable Jet Output Characteristics

The output characteristics of a typical interruptable jet are shown in Figure 6.13. Note that the device has characteristics similar to the jet-interaction amplifier because it also involves the projection and recovery of a free jet.

Vortex Rate Sensor Output Characteristics

The output characteristics of a typical vortex rate sensor are shown in Figure 6.14. Note that the parameter is the rate of turn of the sensor. Note also that because of the type of pickoff used, the characteristics are very nearly straight lines.

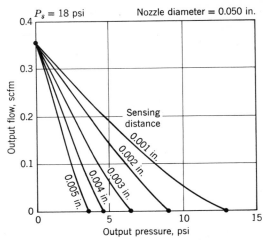

Figure 6.12 Static output characteristics of back-pressure nozzle.

6.6 ACTUATORS

Actuators are defined as the end loads on the fluidic control system that convert fluidic power into mechanical power. Therefore, we will be concerned only with input characteristics. The most common forms are rectilinear and rotary actuators and fluid motors.

The dynamic characteristics of actuators and reflected loads are extremely important to the design of a stable, high-performance fluidic control system. Actuator and load dynamics are covered in Chapter 8.

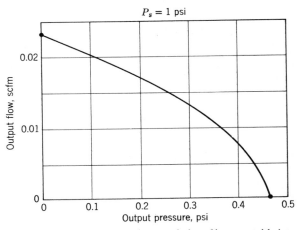

Figure 6.13 Static output characteristics of interruptable jet.

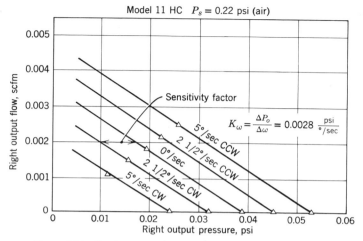

Figure 6.14 Static output characteristics of vortex rate sensor.

Input Characteristics of Rectilinear and Rotary Actuators

Rectilinear and rotary actuators are piston-type devices with very low leakage, designed for limited motion. Therefore, the static input characteristics for a blocked mechanical load, typically as shown in Figure 6.15 are straight lines very nearly on the pressure axis. With no mechanical load, the flow (and velocity) is limited only by the friction and internal resistance of the actuator, but because of limited motion, this condition can exist only for a short period.

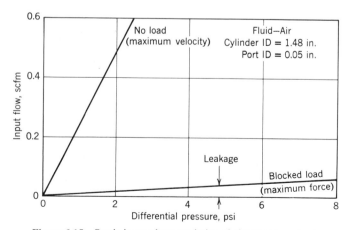

Figure 6.15 Static input characteristics of piston-type actuator.

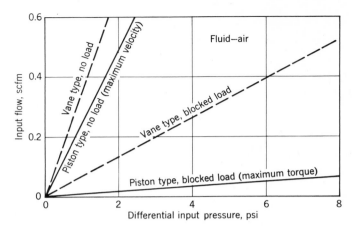

Figure 6.16 Static input characteristics of piston-type and vane-type fluid motors.

Input Characteristics of Fluid Motors

Fluid motors may be piston-type or vane-type devices designed for continuous rotation and may have considerable leakage. Typical static input characteristics with a blocked rotor are shown in Figure 6.16 for a low-leakage piston-type motor and for a high-leakage vane-type motor. When the motors are running at constant velocity, the characteristics are raised parallel to themselves.

7
Large-Signal Performance Analysis

Like most electric and hydraulic components, the characteristics of fluidic devices are nonlinear. When operated at extremes or when driven by a large signal, the performance parameters cannot be considered constant, and the output will be a distorted reproduction of the input signal. In the cases where the effects of these nonlinearities cannot be neglected and the effect of time *can* be neglected, the graphical method of performance analysis is most convenient. Using this method, the system designer can account for the nonlinearities without the need for complex mathematical processes.

Whether or not the effects of nonlinearities can be neglected may be determined from a brief preliminary analysis. The designer must first estimate the magnitude of the signal excursion and how much the output characteristics deviate from linearity over this excursion. He must also estimate values for the time-dependent circuit parameters and how much error they introduce at the expected signal frequencies.

7.1 THE LOAD LINE

Consider first how the jet-interaction amplifier behaves in a circuit. Figure 7.1 illustrates the analogy between the fluidic amplifier and the transistor, and between the fluidic amplifier and the spool valve. Note that it is directly equivalent to a differential connection of transistors and to a three-way spool valve.

Consider now the case of the differential analog fluidic amplifier with a passive load having characteristics as shown in Figure 7.2b. The output

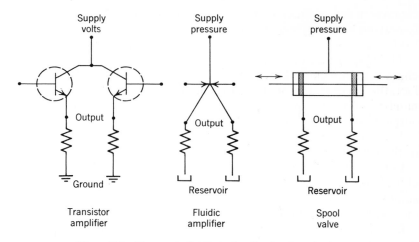

Figure 7.1 Electronic-fluidic-hydraulic circuit analogies.

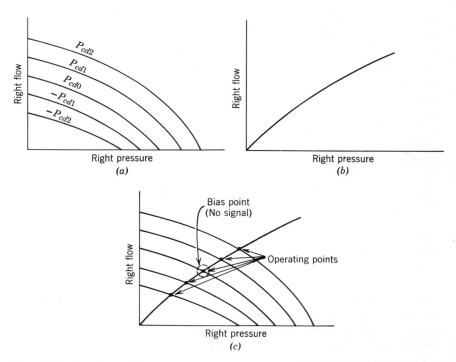

Figure 7.2 Coupling a differential fluidic amplifier to a passive load (*a*) Amplifier output characteristics (right half). (*b*) Passive load characteristics. (*c*) Superposition of characteristics.

characteristics of the amplifier are shown in Figure 7.2a. The problem is to find out what will happen when the amplifier is connected to the passive load.

The first point that the circuit designer must realize is that the output characteristics show how the amplifier will behave *with any load*. In fact, the curves were plotted from the performance of the amplifier for a number of loads, from open outlet port (zero impedance) to blocked output port (infinite impedance).

The second point is that the load characteristic is a single line; that is, there is *only one flow level for each pressure applied*.

Finally, one must realize that when the load is connected to the amplifier, the *pressures and flows are common to the two*; that is, the amplifier output pressure is identical with the passive load pressure and the amplifier output flow is identical with the passive load flow.

Because of these points, the combined behavior of an amplifier with passive load can be found simply by plotting their characteristics on the same graph as shown in Figure 7.2c. The passive load characteristics are superimposed on the amplifier output characteristics as a "load line." Since pressures and flows *must* be identical in both components, the points of *intersection* of the curves can be the *only* operating points.

Consider the case of cascading two differential fluidic amplifiers. For the driving amplifier we would have a set of output characteristics as shown in Figure 7.3a. For the driven amplifier we would be concerned with its input characteristics as shown in Figure 7.3b. Now how can we tell how the amplifiers will behave when they are connected together?

When the output of the driver is connected to the input of the driven amplifier, the output pressure and flow of the driver *must* be identical with the input pressure and flow of the driven. That is, the only *possible* operating conditions are those where the output pressure and flow of the driver amplifier coincide with the input pressure and flow of the driven amplifier. These points are easily found by superimposing the input characteristics of the driven amplifier as a "load line" on the output characteristics of the driver, as shown in Figure 7.3. The points where the characteristics intersect are the only stable operating points. (Note that even in the case of a differential connection, it is necessary to consider each side separately.) Once the range of operating points is defined on the input characteristic, the points may be entered on the output characteristics of the second stage, and the output of the second stage is then determined by the same "load line" method.

Digital devices and circuits can be analyzed using static characteristics in a similar way as illustrated in Figure 7.4. Again the input characteristics of the driven stage (or stages) are superimposed on the output charac-

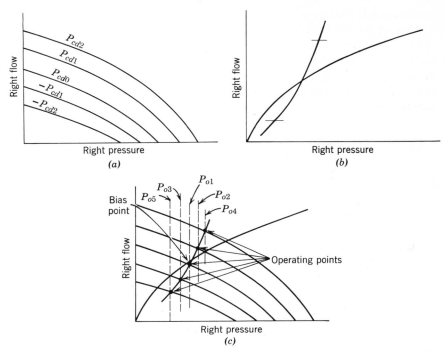

Figure 7.3 Coupling two differential fluidic amplifiers. (*a*) First stage output characteristics (right half). (*b*) Second stage input characteristics. (*c*) Superposition of characteristics.

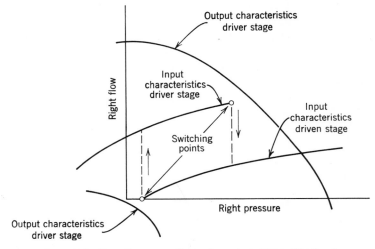

Figure 7.4 Coupling two wall-attachment amplifiers (flip-flops).

teristics of the driving stage. And the only stable operating points are defined by the intersections of these characteristic curves.

The load line concept can be generalized as follows. *Whenever two fluidic components are connected together, the coupled behavior can be defined by superimposing the appropriate characteristic curves for the two components.* The only stable operating points are where the characteristics intersect. The process is illustrated in Figure 7.5.

Load Line Design Procedure

To determine the characteristics of two coupled active or passive fluidic components, two sets of information must be obtained. They are (a) the output characteristics of the driving component (with control variable as a parameter) and (b) the input characteristics of the driven component (with normal operating range marked on the curves).

The procedure is as follows:

1. Plot the driving amplifier output characteristics to a convenient scale.
2. Convert the input characteristics of the driven component to a similar scale.

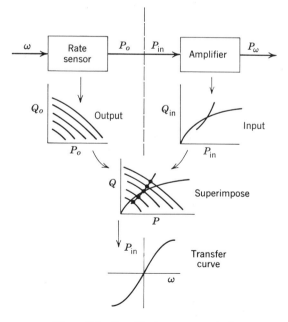

Figure 7.5 Graphical systems analysis.

120   Large-Signal Performance Analysis

3. Plot the input characteristics of the driven component on the output characteristics of the driving component.
4. Note the points at which the curves intersect.
5. The intersecting points determine the operating characteristics of the two coupled fluidic components.

7.2 CALCULATION OF THE TRANSFER (GAIN) CURVE

Once the operating conditions have been defined by the superposition of characteristic curves, the static gain (or performance) curve can be calculated.

Referring again to Figures 7.2 and 7.3 it is first necessary to determine the bias (or quiescent) point. This is given at the intersection of the zero control curve of the driver amplifier with the passive load characteristic or the bias curve of the driven amplifier. This is the point at which the pressure and flow will be *when there is no signal into the driver amplifier*.

When the differential amplifier receives an input signal, one output port pressure increases while the other output port pressure decreases. Since this is the condition that is applied at the input of our driven amplifier, it is appropriate to use the differential curve for the "incremental" load line, that is, for changes about the operating bias point.

To plot the differential pressure gain curve for the driver amplifier loaded with the second differential amplifier, the coordinates shown in Figure 7.6 are used. Where $P_{cd} = 0$, the output pressure of the right port is P_{o1} and the output pressure of the left port (if the amplifier is perfectly balanced) is P_{o1}. Therefore, the differential output P_{od} is zero. When $P_{cd} = +1$, the right output is P_{o2}, the left output P_{o3} and the difference $P_{od} = +2$. When $P_{cd} = -1$, the right output is P_{o3}, the left output is P_{o2}, and the difference is $P_{od} = -2$. Continuing this procedure of taking increments of P_{cd} and calculating the value of P_{od} from the curves leads to a complete transfer (gain) curve as shown in Figure 7.6. Note that this defines the pressure gain of the *driver* amplifier only when it has the driven amplifier as a load.

If the amplifier is not perfectly balanced, different output characteristics for the right output and the left output are used in a way similar to that above.

Transfer Curve Design Procedure

To determine the sensitivity or gain of coupled fluidic components, the plot of output characteristics of the driving component with the input

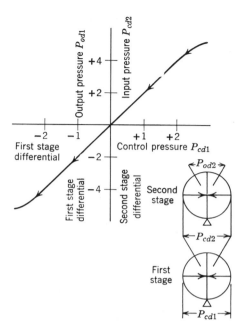

Figure 7.6 Pressure transfer curve of differential amplifier loaded with second differential stage.

characteristics of the driven component superimposed must be available. The procedure is as follows:

1. Select the input and output variables of interest (pressure, flow, rate of turn, etc.).
2. Prepare a graph with scales to conveniently cover the range of variables.
3. Find the no-signal operating (bias) point from the intersection of the zero-control output curve with the load line.
4. Take a number of positive and negative increments of input signal, and determine from the intersection of the appropriate output curves with the load line a number of points along the transfer curve.
5. Plot the results on the previously prepared graph.
6. The resulting transfer curve defines the specific performance of the coupled components under the given conditions of supply bias and load.
7. The slope of the transfer curve is the gain (of an amplifier) or sensitivity (of a transducer).

7.3 STATIC MATCHING OF CASCADED FLUIDIC COMPONENTS

In the preceding section, the load line method for determining the performance of cascaded fluidic components was introduced. In the illustration given (Figure 7.3) the ideal case was assumed, that is, no matching problems arose. In this section the more probable situation is covered when matching problems do arise.

Objectives

The objectives in properly cascading fluidic components are the following:

1. Providing proper gains (or functions).
2. Matching operating bias points.
3. Matching operating ranges.

Let us first define each objective in more detail.

Proper gains are, after all, what the analog circuit is usually there to provide. However, the designer may want primarily flow gain, considering pressure gain and power gain secondary. Or he may want to optimize pressure gain instead.

Proper functions are usually what the digital circuit is to provide. However, sometimes a device may be required to fan out the signal to drive a number of other devices in parallel. Or the designer may want to provide a simple power gain alone.

Operating bias (quiescent) points are the pressures and flows defining the desired conditions in the component with no signal applied. Operating ranges are the ranges of pressures and flows over which the component can be operated with good results.

Matching a Vortex Rate Sensor and a Differential Amplifier

Now consider the matching problem in more detail. Suppose we are given an existing vortex rate sensor to measure rates of turn from 10°/sec counterclockwise to 10°/sec clockwise. Our task is to amplify the output using available "off-the shelf" amplifiers.

Vortex rate sensors are inherently low pressure, high output impedance devices. Typical output characteristics are shown in Figure 7.7. Although push-pull differential circuits are to be used in this application, it is necessary to match the operating characteristics of each half; therefore, we will be concerned with single-ended characteristics in the matching process.

Vented jet-interaction amplifiers are available in a limited number of

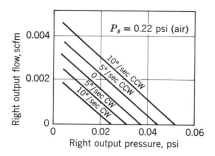

Figure 7.7 Static output characteristics of vortex rate sensor.

standard sizes. An amplifier of high input impedance is necessary to match the high output impedance of the rate sensor. This implies an amplifier with small control nozzles, therefore an amplifier of small overall size. Figure 7.8 shows the input characteristics of a typical small jet-interaction amplifier with power nozzle 0.010×0.025 inches. Note that the preferred bias operating point is 10% of supply pressure, and the linear range of amplification is about $\pm 5\%$ of supply pressure.

Following the procedures for determining the operating characteristics of the rate sensor and amplifier when they are connected together, the input characteristics of the amplifier are superimposed as a load line on the output characteristics of the rate sensor, as shown in Figure 7.9.

Providing Proper Gains

Again referring to Figure 7.9, it is apparent that taking increments of rate of turn to plot the transfer curve yields relatively small increments of input pressure. This condition is evidently because the input characteristic of the amplifier is relatively steep when compared with the rate sensor output characteristics; that is, the impedance match is poor.

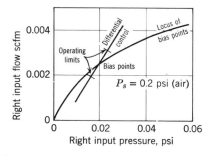

Figure 7.8 Static input characteristics of small vented jet-interaction amplifier.

124 Large-Signal Performance Analysis

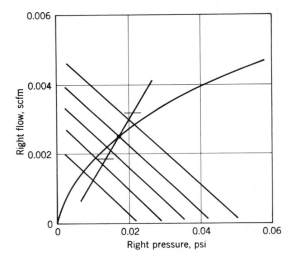

Figure 7.9 Superposition of static characteristics of vortex rate sensor and vented jet-interaction amplifier.

Now suppose it were necessary to optimize *pressure* sensitivity of the amplifier rate-sensor circuit. It is apparent that we would require an amplifier input characteristic with relatively low slope (high impedance) as illustrated in Figure 7.10. Then when increments of rate of turn were

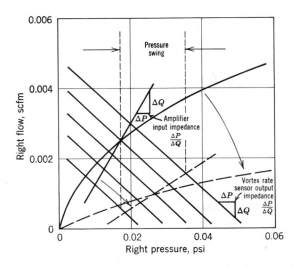

Figure 7.10 Impedance matching for high static pressure gain.

taken to determine the resulting increments of amplifier input pressure, the pressure sensitivity would be vastly increased.

Matching Operating Bias Points

Since the preferred bias point for the amplifier does not coincide with the zero rate of turn curve of the rate sensor, Figure 7.9 illustrates a case of mismatch of the bias points. There are at least three ways to be explored to correct the situation. The output bias level of the rate sensor can be increased, as in Figure 7.11. Or the amplifier supply pressure can be reduced, maintaining the input bias at 10% of the supply, as in Figure 7.12. Or the effective load line of the rate sensor can be shifted by the addition of restrictors in series or in parallel with the amplifier input, as in Figure 7.13. (The latter method is obviously not suitable for correcting the type of mismatch illustrated in this example problem.)

Matching Operating Ranges

With reference to Figure 7.9, it is apparent that there is also a mismatch of optimum operating ranges. The rate sensor, in the situation illustrated, is capable of overdriving the amplifier into its nonlinear range. Again, there are at least three methods to be investigated for correcting the situation: adding series or shunt resistance in the *differential circuit*; changing the output bias of the rate sensor; and changing the amplifier

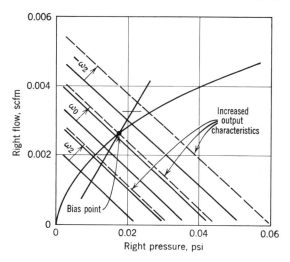

Figure 7.11 Matching operating points by raising rate sensor output bias.

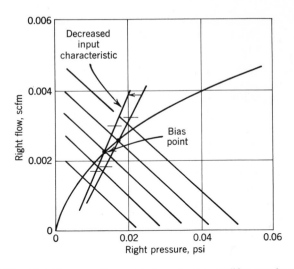

Figure 7.12 Matching operating points by reducing amplifier supply pressure.

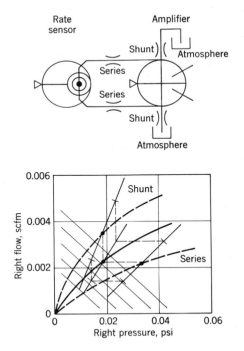

Figure 7.13 Matching operating points by adding restrictors in each side of the circuit.

supply pressure. *Note that resistances across the differential lines will affect the slope and length of the differential load line but not the operating point.* (See Figure 7.14.)

It is evident that two of these steps are also used to match the operating points, and therefore the effect of one upon the other must be considered. It turns out that static matching of fluidic components is a series of compromises based on a thorough understanding of their behavior.

Matching a Digital Amplifier with Three Parallel Flip-Flops

The same approach can be used for digital component matching. So long as the switching portion of the input characteristic falls within the boundaries set up by the two output characteristic curves of the driving component, the devices are roughly matched; that is, the driving amplifier is capable of switching the driven amplifier. However, if it is required to get maximum efficiency; that is, match the range capability of the driving amplifier with the range capability of the driven amplifier, then it is necessary to match the input and output characteristic curves.

Suppose we are given the task of designing a circuit for properly matching a digital amplifier with three parallel-connected flip-flops, all of similar

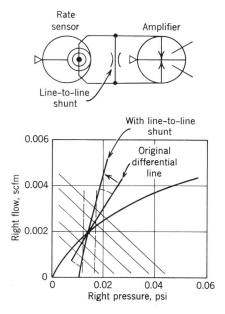

Figure 7.14 Matching operating ranges by adding restrictor between differential lines.

design. The problem is to ensure that the coupled components will be operated in their most efficient mode.

Typically digital amplifiers are used to raise the power level of a signal; therefore, they are designed to be relatively powerful devices. The output characteristics of an available digital amplifier are shown in Figure 7.15. Note that this amplifier is designed to work well with a below-atmospheric pressure at the output port, as indicated by the "off" characteristic curve in the negative-pressure region.

Fluidic flip-flops are made in a wide range of sizes and most are based on the wall-attachment operating principle. Because in this application they are to be used as logic devices only, so we will choose a very small unit and minimize the power consumption of the circuit. The input characteristics are shown in Figure 7.16.

Note that this curve was taken with the bias at zero gage pressure (atmospheric). Increasing the bias pressure within reasonable limits simply shifts the curve upward and to the right.

Following the procedures for determining the characteristics of the digital amplifier driving the three parallel flip-flops, the input characteristics are superimposed onto the output characteristics of the digital amplifier as shown in Figure 7.17. Note that the effective load on the digital amplifier is made up of the individual input characteristics of the three parallel-connected flip-flops, essentially just $\frac{1}{3}$ of the impedance of one, determined graphically by tripling the flow of one unit at every pressure level.

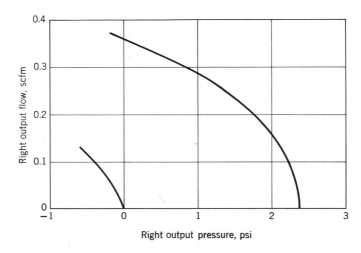

Figure 7.15 Static output characteristics of available digital amplifier.

Static Matching of Cascaded Components   129

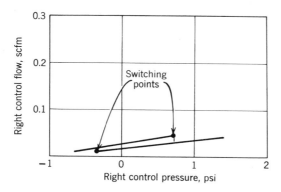

Figure 7.16 Static input characteristics of small flip-flop.

Matching Operating Bias Points

Since the switching limits of the flip-flops are not symmetrically located with respect to the output limits of the digital amplifier Figure 7.17 represents a case of a mismatch of bias points. There are at least three ways to be explored to correct this mismatch. The output bias level of the digital amplifier could be reduced by reducing the supply pressure to it as shown in Figure 7.18. (This is obviously not a solution to the problem.) The input bias level of the flip-flops could be raised by raising the pressure on the opposite control nozzle as shown in Figure 7.19. Or the effective

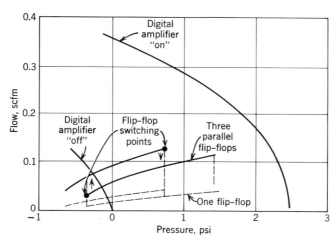

Figure 7.17 Superposition of static characteristics of digital amplifier and three parallel flip-flops.

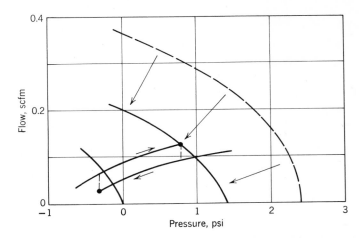

Figure 7.18 Matching operating points by reducing digital amplifier supply pressure.

load on the digital amplifier could be increased by adding restrictors as shown in Figure 7.20.

Matching Operating Ranges

Since the switch points of the flip-flops do not coincide with the output limits of the digital amplifier, Figure 7.17 also illustrates a mismatch of operating ranges. The amplifier, as shown, is capable of overdriving the

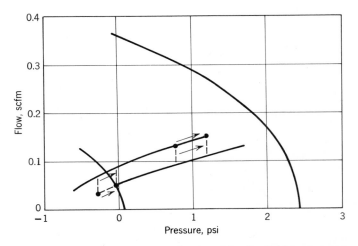

Figure 7.19 Matching operating points by raising flip-flop bias pressure.

Static Matching of Cascaded Components   131

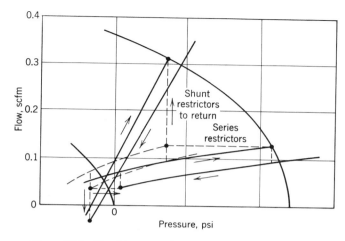

Figure 7.20 Matching operating points by adding restrictors.

flip-flops, seriously affecting the speed of response of the circuit. Again there are three ways to be explored for correcting the mismatch: (1) decreasing the supply pressure to the digital amplifier, (2) increasing the supply pressure to the flip-flops, and (3) adding series or shunt restrictors as illustrated in Figure 7.21.

Again it is clear that matching operating points and operating ranges of digital components also involves a series of compromises based on a thorough understanding of their behavior.

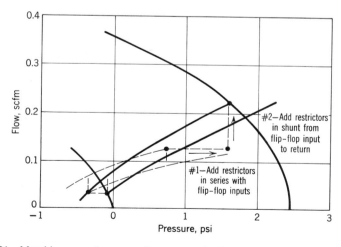

Figure 7.21 Matching operating ranges by proper selection of series and shunt restrictors.

Procedure for Matching

To properly match fluidic components for optimum performance, it is necessary to do the following:

1. Provide proper gain (or function).
2. Match operating bias points.
3. Match operating ranges.

The procedure for doing this is as follows:

1. Obtain a plot of the output of the driving component with preferred operating bias points and operating ranges marked.
2. Choose a driven component whose input characteristics have a slope (impedance) that will optimize the gain variable of interest (pressure, flow, or power).
3. Match operating bias points by adjusting output pressure levels, changing supply pressures, or connecting passive restrictors in series or in parallel with the single-ended circuit.
4. Match operating ranges by changing supply pressures, adjusting bias points, or connecting passive restrictors in series or in parallel with the differential circuit.

8

Equivalent Circuits for Typical Fluidic Devices

Graphic methods of performance analysis are truly general, and if complete data are available, they are valid in all situations. However, for small signals, the reading of graphs becomes inaccurate. In this case a more exact and convenient method of calculating performance is by linearizing parameters around the operating bias point and employing them in an equivalent electrical circuit. This approach is widely used in all forms of engineering analysis, including electronics, acoustics, pneumatics, hydraulics, and mechanics. Therefore, by applying this approach in fluidics, all the mathematical tools developed for those forms over so many years can also be used to advantage in the analysis of fluidic systems.

In digital circuit analysis, the equivalent electric circuit is also of value, in spite of the fact that linearizing and lumping parameters for such large-signal excursions represents gross oversimplification. Specifically the electrical circuit is of value to the digital circuit designer in providing considerable insight into the dynamics of the circuit and in providing the basis for a rough estimate of transient response.

8.1 ANALOG FLUIDIC AMPLIFIERS

The process of developing an equivalent electrical circuit for a fluidic component can be a difficult analytical task. Fortunately, useful mathematical models can also be developed through a process of logic and

134   Equivalent Circuits for Fluidic Devices

validated through comprehensive experimental tests. To date, this has been the only known approach that has produced useful results (Refs. 2 and 3).

As an illustration of the approach, we will consider the case of the vented jet-interaction amplifier in some detail. Mainly by a process of logic, a direct lumped model of the amplifier can be fabricated as shown in Figure 8.1 using Z to represent fluid impedance. Each element in the circuit represents a measurable fluid phenomenon in the amplifier, and each is nonlinear. If each nonlinearity were fully described, the circuit could be used for large-signal performance analysis. However, because they have not yet been fully described, the circuit of Figure 8.1 is most useful for the insight it provides into the factors involved in cascading amplifiers and as the basis for deriving the simplified linearized equivalent circuit for small-signal performance analysis.

The general layout is directly analogous to the actual device. The power supply is a pressure source developing pressure P_S and having an internal

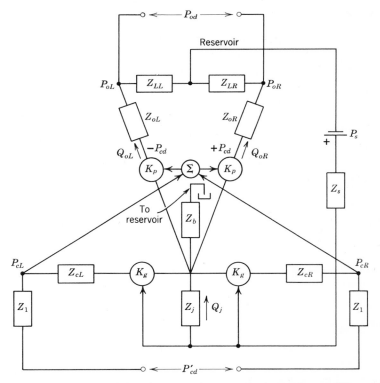

Figure 8.1   Lumped electrical model for vented jet-interaction amplifier.

impedance Z_S. The pressure appears at the supply port of the amplifier, causing a power jet flow Q_j through the power jet impedance Z_j and the branch and load impedances Z_{oL}, Z_{oR}, Z_{LL}, and Z_{LR}.

In each branch there is a generator with a gain K_p, the calculated amplification factor. Z_{oL} and Z_{oR} are the reciprocals of the slopes of the load circuit characteristics at the operating point. Z_{LL} and Z_{LR} are the reciprocals of the slopes of the load restrictor characteristics at the operating point. Z_B is the effective impedance of the bleed ports. The generators are connected differentially and driven by a function of the differential control pressure P_{cd}. Differential output pressure P_{od} is developed at the amplifier output ports.

In the control circuit we have the internal control circuit impedance for each side, and Z_1, representing any series impedance in the control signal inlet. There are also zero impedance voltage generators to simulate the suction pressure in the interaction region, generated by the power jet flow Q_j, having a gain of K_g. Finally, the absolute levels of control pressure P_{CL} and P_{CR} are applied at the control terminals after the signal has passed through the line impedance Z_1. The effects of internal feedback are neglected.

Once the nonlinear model has been hypothesized, the next step is to generate the linearized equivalent circuit for small signals. Referring to the complete circuit of Figure 8.1 and using Δ to denote incremental changes, we have in the left branch of the load circuit

$$P_S = Q_j(Z_S + Z_j) - K_p P_{cd} + (Q_{oL} - \Delta Q_{oL})(Z_{oL} + Z_{LL}) \qquad (1)$$

and in the right branch

$$P_S = Q_j(Z_S + Z_j) + K_p P_{cd} + (Q_{oR} + \Delta Q_{oR})(Z_{oR} + Z_{LR}) \qquad (2)$$

Subtracting eq. 1 from eq. 2 to eliminate the steady conditions and assuming

$$Z_{oL} = Z_{oR} = Z_o \quad \text{and} \quad Z_{LL} = Z_{LR} = Z_L \quad \text{and} \quad \Delta Q_{oL} = \Delta Q_{oR} = \Delta Q_o$$

we have

$$2K_p P_{cd} = \Delta Q_o(2Z_o + 2Z_L) \qquad (3)$$

This represents an equivalent electrical circuit with a generator of output $2K_p P_{cd}$, an internal impedance of $2Z_o$ and a load of $2Z_L$.

Similarly in the control circuits we have

$$P_{cL} + \Delta P_{cL} = (Q_{cL} + \Delta Q_{cL})Z_{cL} - (Q_j + \Delta Q_j)Z_j K_g \qquad (4)$$

and

$$P_{cR} - \Delta P_{cR} = (Q_{cR} - \Delta Q_{cR})Z_{cR} - (Q_j + \Delta Q_j)Z_j K_g \qquad (5)$$

Subtracting eq. 5 from 4 to eliminate steady conditions and assuming

$$\Delta P_{cL} = \Delta P_{cR} = \Delta P_c \quad \text{and} \quad \Delta Q_{cL} = \Delta Q_{cR} = \Delta Q_c \quad \text{and} \quad Z_{cL} = Z_{cR} = Z_c$$

we have

$$2\Delta P_c = 2\Delta Q_c Z_c \tag{6}$$

If we define the differential pressure $P_{cd} = 2\Delta P_c$, then equation 6 expresses the conditions in an electrical circuit with a generator of P_{cd} and an impedance $2Z_c$.

Thus the *linearized* incremental equivalent circuit can be drawn as illustrated in Figure 8.2. Note that the equivalent circuit eliminates bias conditions, supply pressure, and, in this case, bleed impedance. Note also that the equivalent circuit elements are now treated as *constants*, calculated for incremental changes around the operating bias point. The load circuit equivalent generator (representing pure amplifier gain) is represented by $2K_p$; the impedance in series with the output is $2Z_o$ and the effective load impedance between output ports is $2Z_L$. The input circuit, neglecting any impedance in the line leading to the control nozzles, is represented as a shunt impedance $2Z_c$.

The equivalent circuit of Figure 8.2 is valid for small-signal analysis at low frequencies, providing that the impedances are properly defined at the frequencies of interest. If only the "resistive" components of the impedance are used, the circuit is good for the static case only.

Equivalent Circuit for a Vented Jet-Interaction Amplifier (Refs. 4 and 5)

At higher frequencies where resistive elements no longer satisfactorily describe the behavior, the time delays due to transit time, wave propagation, and the presence of "reactive" circuit elements (like volume

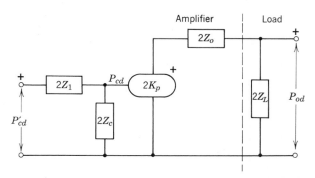

Figure 8.2 Linearized small-signal equivalent circuit for vented jet-interaction amplifier (valid for static and low frequency cases).

capacitance) must be considered. Figure 8.3 defines the equivalent electrical circuit for higher frequencies. The element in series with the input circuit $2L_c$ is due to inertance in the line to the control nozzle. The shunt elements $2R_c$ and $C_c/2$ are effective control nozzle resistance and volume capacitance of the control line. The equivalent generator $2K_p$ contains a delay factor (e^{-st_d}) which includes wave propagation and transit times in the total path from the control port to the load terminals. The output circuit contains a series inductor $2L_o$ and resistor $2R_o$ and a shunt volume capacitor $C_o/2$. If the lines to the load are short, the load volume capacitance $C_L/2$ is directly parallel with the amplifier capacitance, and the load resistance $2R_L$ parallels both. The transfer function for this amplifier is

$$\frac{P_{od}}{P'_{cd}} = \frac{2K_p R_L}{R_o + R_L}$$

$$\times \frac{e^{-st_d}}{\left\{\left(1 + s\frac{L_c}{R_c} + s^2 L_c C_c\right)\left[1 + s\left(\frac{C_t 2 R_o R_L}{R_o + R_L} + \frac{L_o}{R_o + R_L}\right) + s^2 \frac{2 C_t L_o R_L}{R_o + R_L}\right]\right\}}$$

The overall high-frequency transfer function from Figure 8.3 contains an attenuation due to the output circuit resistor network, a gain factor equal to twice the amplification factor, the time delay, and quadratic factors resulting from the combination of time constants in the input and output networks.

The equivalent circuit for the vented jet-interaction amplifier has been proven by experiment to be valid to frequencies above 400 Hz.

Equivalent Circuit for a Closed Jet-Interaction Amplifier (Ref. 3)

The equivalent circuit for the closed jet-interaction amplifier is shown in Figure 8.4. It is similar to the circuit for the vented amplifier except that

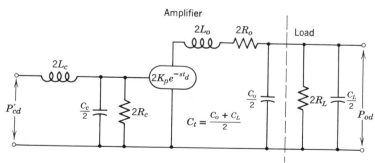

Figure 8.3 Equivalent circuit for vented jet-interaction amplifier valid to 400 Hz.

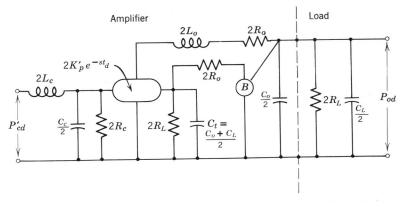

Figure 8.4 Equivalent circuit for closed jet-interaction amplifier valid to 400 Hz.

it includes an internal feedback loop. The feedback loop is present as a result of the backup of output pressures into the interaction chamber, where they act to reduce the gain of the amplifier circuit. The transfer function for this amplifier is

$$\frac{P_{od}}{P'_{cd}} \cong \frac{2K_p R_L}{R_o + R_L} \frac{\left(1 + sC_t \frac{2R_o R_L}{R_o + R_L}\right) e^{-st_d}}{\left(1 + s\frac{L_c}{R_c} + s^2 L_c C_c\right)\left[1 + s\frac{4C_t R_o R_L}{(1 + 2K'_p B)(R_o + R_L)}\right.}$$

$$\left. + s^2 \frac{4C_t^2 R_o^2 R_L^2}{(1 + 2K'_p B)(R_o + R_L)^2} + s^3 \frac{4L_o C_t^2 R_L^2 R_o}{(1 + 2K'_p B)(R_o + R_L)^2}\right]$$

The equivalent circuit for the closed jet-interaction amplifier has been validated by experiment to frequencies above 400 Hz.

Equivalent Circuit for Vented Elbow Amplifier (Ref. 3)

The equivalent circuit for a typical vented elbow amplifier is shown in Figure 8.5. Note that this amplifier is essentially a flow amplifier; hence, the equivalent generator is a flow generator. Otherwise, the circuit is similar in form to the jet-interaction amplifier circuits. The transfer function for this amplifier is

$$\frac{P_o}{P_c} = \frac{K_f R' R_L}{R_c (R_o + R_L)}$$

$$\times \frac{e^{-st_d}}{\left\{\left(1 + s\frac{L_c}{R_c} + s^2 L_c C_c\right)(1 + sC'R')\left[1 + s\frac{C_t R_o R_L + L_o}{R_o + R_L} + s^2 \frac{L_o C_t R_L}{R_o + R_L}\right]\right\}}$$

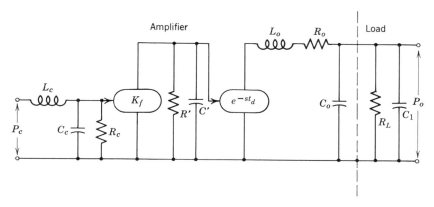

Figure 8.5 Equivalent circuit for vented elbow amplifier valid to 400 Hz.

The equivalent circuit for the vented elbow amplifier has been validated by experiment to frequencies above 400 Hz.

8.2 DIGITAL FLUIDIC AMPLIFIERS

Equivalent Circuit for Vented Wall-Attachment Amplifier

The equivalent circuit for digital vented wall-attachment amplifier is shown in Figure 8.6. The element in series with the input circuit is the effective inductance $2L_c$ due to inertance in the line. The shunt elements $2R_c$ and $C_c/2$ are the effective control nozzle resistance and the effective volume capacitance. These elements are at least double-valued, depend-

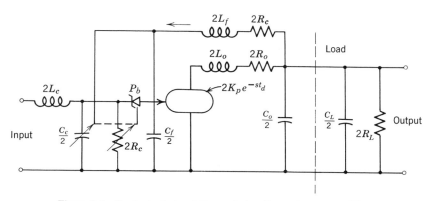

Figure 8.6 Equivalent circuit for vented wall-attachment amplifier.

140 Equivalent Circuits for Fluidic Devices

ing on the state of the output circuit. Therefore there is a feedback loop that contains some dynamics due to conditions in the interaction region, which switches the effective input impedances to different values when the power stream switches from one wall to the other.

The equivalent generator $2K_p$ is effectively a pressure switch triggered at a level determined by feedback-controlled reference diodes. It contains a delay factor (e^{-st_d}) that includes wave propagation and transit times in the total path from the control port to the output ports. The output circuit is similar to the one for the proportional amplifier, containing series inductance, shunt capacitance, and series resistance. If the loads are closely coupled the load impedances are directly in parallel with the amplifier capacitance.

The dynamic response contains a nonlinear second-order term due to the input circuit, a time delay due to transit time, and a "linear" second-order term due to the output circuit.

8.3 FLUIDIC SENSORS

Equivalent Circuit for a Vortex Rate Sensor (Refs. 2 and 4)

The equivalent circuit for a typical vortex rate sensor is shown in Figure 8.7. The rate of turn drives an equivalent generator containing the sensitivity factor. In the process it is subject to a time delay due to the time that it takes for the angular momentum at the rim to be transported through the vortex to the pickup in the drain tube. The resulting pressure signal is then applied to an output circuit containing the same arrangement of elements that appear in the amplifier output circuits. The transfer

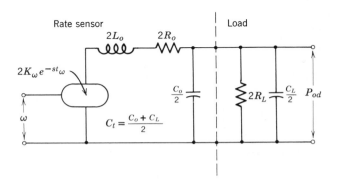

Figure 8.7 Equivalent circuit for vortex rate sensor.

function for the vortex rate sensor is

$$\frac{P_{od}}{\omega} = \frac{2K_\omega R_L}{R_o + R_L} \frac{e^{-st_\omega}}{1 + s\left(\frac{2C_t R_o R_L}{R_o + R_L} + \frac{L_o}{R_o + R_L}\right) + s^2 \frac{2C_t L_o R_L}{R_o + R_L}}$$

8.4 ACTUATORS

Equivalent Circuit for a Piston-Type Actuator

The equivalent electrical circuit for a typical piston-type (or bellows-type) actuator is shown in Figure 8.8. Pressure applied to the actuator causes flow in the equivalent capacitor because of compressibility, and flow through the leakage resistance (if present). A reduced pressure is also applied to a series network containing the reflected acceleration, velocity, and position characteristics of the load. All of these flows must be conducted through the internal resistance of the actuator and its associated connections that limit the maximum available velocity and acceleration of the actuator-load combination (Ref. 7).

Equivalent Circuit for Piston-Type and Vane-Type Motors

The equivalent circuit for a typical piston-type or vane-type motor is shown in Figure 8.9. Pressure applied to the motor causes flow due to

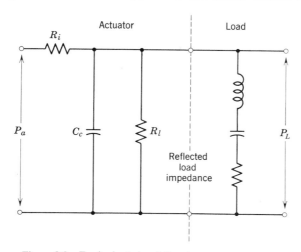

Figure 8.8 Equivalent circuit for piston-type actuator.

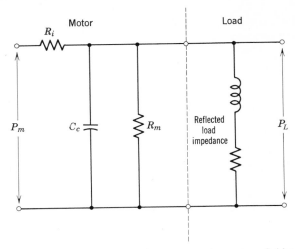

Figure 8.9 Equivalent circuit for piston-type and vane-type fluid motors.

compressibility of the trapped fluid and due to leakage (which is considerable in the vane-type motor). A reduced pressure is also applied to a series network containing the reflected acceleration and velocity characteristics of the load. (The resulting flows through the internal resistance of the motor and its associated connections cause a pressure drop that limits the top speed and acceleration of the motor and its load.)

9

Small-Signal Performance Analysis

When fluidic components are cascaded (i.e., one becomes the load on the other), their equivalent electrical circuits will be cascaded in a similar way. Figure 9.1 illustrates the connection of a vented jet-interaction amplifier to the output of a vortex rate sensor for the purpose of amplifying the signal. The cascading of the equivalent circuits simply involves the connection of the output terminals of the rate sensor circuit (Figure 8.7) to the input terminals of the amplifier (Figure 8.3). Note that in the interconnecting lines we have added a resistance R_1 to represent any additional resistance due to unusual line lengths.

9.1 DERIVATION OF THE TRANSFER FUNCTION FOR CASCADED FLUIDIC COMPONENTS

The derivation of the transfer function for the cascaded fluidic components involves the straightforward analysis of the equivalent electrical circuits by well-known mathematical methods (Ref. 12). Specifically, it involves the loop analysis of each set of *coupled* circuits using the Laplace transform notation, and the combination of the results into a single transfer function representing the overall dynamic response of the cascaded fluidic components. The procedure is perhaps best communicated by means of an example.

Consider the circuit illustrated in Figure 9.1. We must first generate the array of loop equations in the form of

$$P_1 = Z_{11}Q_1 + Z_{12}Q_2 + Z_{13}Q_3$$

144   Small-Signal Performance Analysis

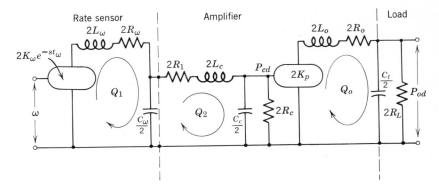

Figure 9.1 Cascading equivalent circuits for vortex rate sensor and vented jet-interaction amplifier.

$$P_2 = Z_{21}Q_1 + Z_{22}Q_2 + Z_{23}Q_3$$
$$P_3 = Z_{31}Q_1 + Z_{32}Q_2 + Z_{33}Q_3$$

In this case we have only two *coupled* loops: those involving Q_1 and Q_2. From these we are to derive the transfer function for the first coupled section P_{cd}/ω. The loop equations, assuming zero initial conditions, are

$$2K_\omega e^{-st_\omega}\omega = \left(2R_\omega + 2sL_\omega + \frac{2}{sC_\omega}\right)Q_1 + \left(\frac{2}{sC_\omega}\right)(-Q_2)$$

$$0 = \left(\frac{2}{sC_\omega}\right)(-Q_1) + \left(\frac{2}{sC_\omega} + 2sL_c + 2R_1 + \frac{\dfrac{4R_c}{sC_c}}{2R_c + \dfrac{2}{sC_c}}\right)(Q_2)$$

Solving for Q_2 by determinants, we have

$$Q_2 = \frac{\begin{vmatrix} \left(2R_\omega + 2sL_\omega + \dfrac{2}{sC_\omega}\right) & 2K_\omega e^{-st_\omega}\omega \\ -\left(\dfrac{2}{sC_\omega}\right) & 0 \end{vmatrix}}{D}$$

where

$$D = \begin{vmatrix} \left(2R_\omega + 2sL_\omega + \dfrac{2}{sC_\omega}\right) & -\left(\dfrac{2}{sC_\omega}\right) \\ -\left(\dfrac{2}{sC_\omega}\right) & \left(\dfrac{2}{sC_\omega} + 2sL_c + 2R_1 + \dfrac{\dfrac{4R_c}{sC_c}}{2R_c + \dfrac{2}{sC_c}}\right) \end{vmatrix}$$

Derivation of the Transfer Function

Then

$$Q_2 = \frac{K_\omega e^{-st_\omega \omega}\left(\frac{1}{sC_\omega}\right)}{\left(R_\omega + sL_\omega + \frac{1}{sC_\omega}\right)\left(\frac{1}{sC_\omega} + sL_c + R_1 + \frac{\frac{R_c}{sC_c}}{R_c + \frac{1}{sC_c}}\right) - \left(\frac{1}{sC_\omega}\right)^2}$$

For simplicity in the algebraic solution for Q_2, let us assume that all the inductances L_ω, L_c, and L_o are negligible. One should keep in mind the fact that we can reintroduce these terms into the solution whenever desired simply by adding the appropriate sL term to each of the associated R terms.

Then

$$Q_2 = \frac{K_\omega e^{-st_\omega \omega}\left(\frac{1}{sC_\omega}\right)}{\left(R_\omega + \frac{1}{sC_\omega}\right)\left(\frac{1}{sC_\omega} + R_1 + \frac{\frac{R_c}{sC_c}}{R_c + \frac{1}{sC_c}}\right) - \left(\frac{1}{sC_\omega}\right)^2}$$

We are attempting to define the transfer function for P_{cd}/ω as a first step, but

$$P_{cd} = Q_2 \frac{\frac{2R_c}{sC_c}}{R_c + \frac{1}{sC_c}}$$

Substituting the equation for Q_2 gives

$$\frac{P_{cd}}{\omega} = \frac{2K_\omega e^{-st_\omega}\left(\frac{1}{sC_\omega}\right)\left(\frac{\frac{R_c}{sC_c}}{R_c + \frac{1}{sC_c}}\right)}{\left(R_\omega + \frac{1}{sC_\omega}\right)\left(\frac{1}{sC_\omega} + R_1 + \frac{\frac{R_c}{sC_c}}{R_c + \frac{1}{sC_c}}\right) - \left(\frac{1}{sC_\omega}\right)^2}$$

This expression then is reduced algebraically to

$$\frac{P_{cd}}{\omega} = \frac{K_\omega e^{-st_\omega} R_c}{(R_\omega + R_1 + R_c)}\left[s^2 C_\omega R_\omega C_c R_c \left(\frac{R_1}{R_\omega + R_1 + R_c}\right)\right.$$
$$\left. + sC_c R_c \left(\frac{R_\omega + R_1}{R_\omega + R_1 + R_c}\right) + sC_\omega R_\omega \left(\frac{R_1 + R_c}{R_\omega + R_1 + R_c}\right) + 1\right]^{-1}$$

This defines the transfer function for the first section of the equivalent circuit involving the coupled rate sensor output and amplifier input circuits.

The second section of the complete equivalent circuit involving the flow Q_o is described by the following two equations.

$$Q_o = \frac{K_p e^{-st_o} P_{cd}}{sL_o + R_o + \dfrac{\dfrac{R_L}{sC_t}}{R_L + \dfrac{1}{sC_t}}}$$

and

$$P_{od} = Q_o \frac{\dfrac{2R_L}{sC_t}}{R_L + \dfrac{1}{sC_t}}$$

Combining gives

$$P_{od} = \frac{K_p e^{-st_o} P_{cd} \dfrac{\dfrac{R_L}{sC_t}}{R_L + \dfrac{1}{sC_t}}}{sL_o + R_o + \dfrac{\dfrac{R_L}{sC_t}}{R_L + \dfrac{1}{sC_t}}}$$

Again neglecting L_o for simplicity,

$$\frac{P_{od}}{P_{cd}} = \frac{2K_p e^{-st_o} \dfrac{\dfrac{R_L}{sC_t}}{R_L + \dfrac{1}{sC_t}}}{R_o + \dfrac{\dfrac{R_L}{sC_t}}{R_L + \dfrac{1}{sC_t}}}$$

This expression is reduced algebraically to

$$\frac{P_{od}}{P_{cd}} = \frac{2K_p e^{-st_o} R_L}{R_o + R_L} \left(sC_t \frac{R_o R_L}{R_o + R_L} + 1 \right)^{-1}$$

Now we have the transfer functions P_{cd}/ω and P_{od}/P_{cd}. We want the overall transfer function P_{od}/ω. But

$$\frac{P_{od}}{\omega} = \frac{P_{cd}}{\omega} \times \frac{P_{od}}{P_{cd}}$$

Then

$$\frac{P_{od}}{\omega} = \frac{4K_\omega e^{-st_\omega} R_c K_p e^{-st_o} R_L}{(R_\omega + R_1 + R_c)(R_o + R_L)} \left[sC_t \left(\frac{R_o R_L}{R_o + R_L} \right) + 1 \right]^{-1}$$

$$\left[s^2 C_\omega R_\omega C_c R_c \left(\frac{R_1}{R_\omega + R_1 + R_c} \right) + sC_c R_c \left(\frac{R_\omega + R_1}{R_\omega + R_1 + R_c} \right) \right.$$

$$\left. + sC_\omega R_\omega \left(\frac{R_1 + R_c}{R_\omega + R_1 + R_c} \right) + 1 \right]^{-1}$$

describes the small-signal static and dynamic behavior of the cascaded rate sensor and amplifier. To calculate the behavior in numerical form, it will first be necessary to evaluate each of the equivalent circuit parameters contained in the above transfer function.

9.2 CALCULATION OF EQUIVALENT CIRCUIT PARAMETERS

Vortex Rate Sensor

The performance parameters for the vortex rate sensor are calculated from the graphical output characteristics and the circuit dimensions at the operating point. Starting with the combined rate sensor-amplifier static characteristics as shown in Figure 9.2, it is possible to define the operating bias point at $P_1 Q_1$. From here we can proceed with the calculation of circuit parameters.

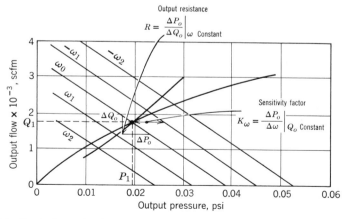

Figure 9.2 Equivalent circuit parameters defined from static output characteristics of vortex rate sensor.

148   Small-Signal Performance Analysis

Sensitivity Factor K_ω. The sensitivity factor is calculated from the horizontal spacing between rate sensor output characteristics at the operating bias point (see Figure 9.2).

Time Delay t_ω. The time delay of the rate sensor is calculated from the dimensions of the rate sensor and output circuit and the flow and pressure conditions at the operating point. It can also be measured directly.

Output Resistance R_ω. The output resistance is calculated from the slope of the zero signal output characteristic curve at the operating point (see Figure 9.2).

Output Inductance L_ω. The output inductance of the rate sensor is calculated from the density at the operating point and the dimensions of the output circuit.

Output Capacitance C_ω. The output capacitance of the rate sensor is calculated from the dimensions of the output circuit and the pressure at the operating point.

Amplifier

The parameters in the amplifier input circuit are calculated from the graphical characteristics that define the conditions in the coupled rate sensor output-amplifier input circuits (see Figure 9.3). The parameters in the amplifier output circuit are calculated from the graphical characteristics that define the conditions in the amplifier output circuit with the resistance R_L as a load line.

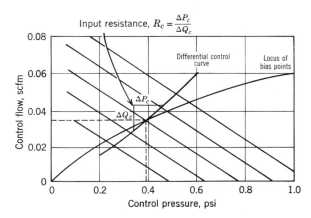

Figure 9.3 Equivalent circuit parameters defined from static input characteristics of jet-interaction amplifier.

Calculation of Equivalent Circuit Parameters   149

Input Inductance L_c. The input inductance of the amplifier is calculated from the dimensions of the input circuit and the control operating point pressure.

Input Resistance R_c. The input resistance of the amplifier is calculated from the slope of the input characteristic at the operating point (see Figure 9.3).

Input Capacitance C_c. The input capacitance of the amplifier is calculated from the dimensions of the input circuit and the control pressure at the operating point.

Line Resistance R_1. The line resistance in the input circuit is calculated from the line dimensions and the pressure and flow conditions at the operating point by using standard formulae or measuring as described in Chapter 5.

Amplification Factor K_p. The pressure amplification factor is calculated from the horizontal spacing between curves of the amplifier output characteristics at the operating point (see Figure 9.4).

Time Delay t_o. The amplifier time delay is calculated from the dimensions of the amplifier circuits and the pressure and flow conditions at the operating point. It can also be measured directly.

Output Resistance R_o. The amplifier output resistance is calculated from the slope of the zero signal output characteristic at the operating point (see Figure 9.4).

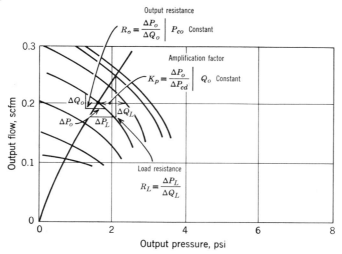

Figure 9.4 Equivalent circuit parameters defined from static output characteristics of amplifier.

150 Small-Signal Performance Analysis

Output Inductance L_o. The amplifier output inductance is calculated from the amplifier output circuit dimensions and the density at the operating point.

Output Capacitance C_t. The output capacitance of the amplifier and its load circuit is calculated from the dimensions of the amplifier and load circuits and the pressure at the operating point.

Load Resistance R_L. The load resistance on the amplifier is calculated from the slope of the load line (load restrictor characteristic) at the operating point (see Figure 9.4).

9.3 CALCULATION OF FREQUENCY RESPONSE

Substitution of the Variable

To calculate the frequency response of cascaded fluidic components, we must first generate the coupled equivalent circuit, derive the transfer function, calculate the performance parameters, and substitute them into the transfer function. The result would be a numerical equivalent of the transfer function containing the Laplace transform variable, s, typically as follows:

$$\frac{P_{od}}{\omega} = 0.05 \frac{e^{-0.01s}}{(8 \times 10^{-6}s^2 + 5 \times 10^{-3}s + 1)(2 \times 10^{-3}s + 1)}$$

With reference to Chestnut and Mayer, *Servomechanisms and Regulating System Design*, Vol. I, one of the properties of the Laplace transform is direct conversion into the frequency plane. That is, the frequency characteristics of the transfer function above can be determined simply by substituting the variable $j2\pi f$ for the Laplace transform variable s. The factor $2\pi f$ is the radian frequency, and $j = \sqrt{-1}$.

Therefore, to determine the frequency response of the above transfer function defining the behavior of the cascaded vortex rate sensor and amplifier, we substitute $j2\pi f$ for s (except in the delay factor $e^{-0.01s}$).

$$\frac{P_{od}}{\omega} = 0.05 \frac{e^{-0.01s}}{(-3.2 \times 10^{-4}f^2 + j3.1 \times 10^{-2}f)(j1.3 \times 10^{-2}f + 1)}$$

Now the calculation must be separated into two parts, one involving the linear system response and one involving the time delay (which we will

Calculation of Frequency Response   151

consider a nonlinear system response); schematically,

(linear)   (nonlinear)

$$\frac{P_{od}}{\omega} = \frac{0.05}{(1 - 3.2 \times 10^{-4} f^2 + j3.1 \times 10^{-2} f)(j1.3 \times 10^{-2} f + 1)} \quad e^{-0.01s}$$

Calculation of Linear Response

For the purpose of illustration, let us calculate the response of the cascaded rate sensor and amplifier at a frequency of 100 Hz.

In the linear portion of the transfer function, the frequency is substituted directly:

$$\frac{P_{od}}{\omega} = \frac{0.05}{(1 - 3.2 \times 10^{-4} \times 10^4 + j3.1 \times 10^{-2} \times 10^2)(j1.3 \times 10^{-2} \times 10^2 + 1)}$$

$$\frac{P_{od}}{\omega} = \frac{0.05}{(-2.2 + j3.1)(j1.3 + 1)}$$

Converting the complex numbers to vectors,

$$\frac{P_{od}}{\omega} = \frac{0.05}{3.8 \underline{/124°} \times 1.6 \underline{/39°}} = \frac{0.05}{6.1 \underline{/163°}}$$

$$\frac{P_{od}}{\omega} = 0.008 \underline{/-163°} \quad \text{at 100 Hz}$$

and converting to decibels,

$$\frac{P_{od}}{\omega} = -42 db \; \underline{/-163°} \quad \text{at 100 Hz.}$$

Calculation of the Response to Time Delay

The factor $e^{-0.01s}$ in the transfer function of the cascaded rate sensor and amplifier represents a dead-time delay: a nonlinear condition. Fortunately, its effect on frequency response is not difficult to analyze.

Consider the illustration of Figure 9.5. It is evident that the effect of dead time is to delay the sine wave a fixed time without affecting the amplitude. However, it is also to be recognized that a fixed delay time represents a different phase shift (in terms of number of degrees), depending on the frequency of the sine wave. In fact, it turns out that the effective phase shift in degrees is a linear function of frequency; that is,

$$\theta = -360 f t_d \text{ degrees}$$

152   Small-Signal Performance Analysis

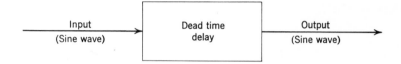

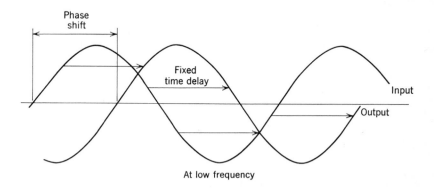

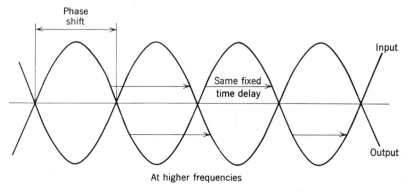

Figure 9.5   Effect of dead time on sinusoidal response.

which is logical because if the frequency is 1 Hz and the time delay is 1 sec, the wave would appear to be shifted 360°.

Thus, with the given time delay of 0.01 sec, the effective phase shift at 100 Hz is

$$\theta = -360 \times 100 \times 0.01$$
$$\theta = -360°$$

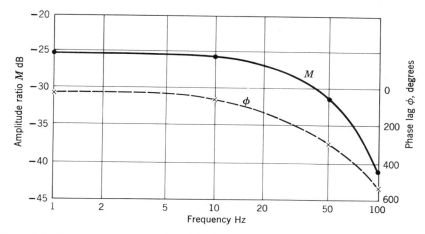

Figure 9.6 Frequency response of combined rate sensor and vented jet-interaction amplifier (Bode diagram).

The Total Frequency Response

Considering the response to both the linear and the nonlinear portions of the transfer function, the value of the complete transfer function is the direct cascade of the response of each alone. That is,

$$\frac{P_{od}}{\omega} = -42db \underline{/-163°} \underline{/-360°}$$

$$\frac{P_{od}}{\omega} = -42db \underline{/-523°} \quad \text{at 100 Hz.}$$

Other values of frequency are inserted into the transfer function to calculate the linear response, and the nonlinear response, which combine to give the total response at each point. The results are plotted as a Bode diagram in Figure 9.6.

10
Detailed Systems Design Procedure

Refer to the appropriate chapter in this book for information related to each of the items of the following.

10.1 REQUIRED INFORMATION

1. Performance specification for the system.
2. Characteristics of available supplies, signal sources, and driven loads.
3. Static characteristics on available fluidic and interface system components (input, output, and power nozzle).
4. Internal dimensional characteristics of available fluidic and interface system components (effective volumes, lengths, and areas).

10.2 STEP-BY-STEP DESIGN

1. Choose final control element suitable for driving the given load (actuator, indicator). The choice is a function of the available supply, the required force and velocity levels, and the function to be performed.
2. Determine its input characteristics (pressure, flow, volume, compliance). This includes the static characteristics, the physical dimensions, and the reflected dynamics of the load.
3. Choose a sensor suitable for detecting the input variable (sensitivity, signal-to-noise ratio). In some cases the input variable may be given as a part of the system specifications.
4. Determine its output characteristics (static characteristics, effective

volumes, lengths, time delay). When the input variable is specified, the static loading characteristics and the physical dimensions must also be specified.

5. Block out the system between sensor (input) and final control element (output) with appropriate functional components (impedance level, pressure gain, power capacity, speed of response, compensation, computation).

6. Tentatively choose fluidic devices with the potential for satisfying the functional requirements of the system. Where standard devices are available, they should be used; where standard devices are not available, their characteristics must be defined.

7. Explore the operating bias point matching problem by superimposing the input characteristics of each driven component onto the output characteristics of each driving component. The operating bias point is at the intersection of the zero control parameter curve with the effective load line.

8. Explore the matching of preferred operating ranges using the same superimposed characteristics. Matching operating ranges is in some ways similar to the operating bias point matching problem, except restrictors can now be connected across the differential circuit without affecting the bias point.

9. Make the operating bias points coincide and the operating ranges compatible by suitable tradeoffs and adjustments (supply pressure, bias level, shunt, and series resistances).

10. Calculate the transfer (or switching) curves of the system in steps beginning from the sensor and including required matched gain stages as needed to provide the necessary signal level at the final control element.

11. Investigate and correct for tendencies toward nonlinearity, saturation, and inefficient use of operating ranges.

12. Select the appropriate equivalent electrical circuit for each component of the system.

13. Prepare an equivalent electrical circuit of the entire coupled system.

14. Derive or estimate the transfer function of each isolated portion of the system equivalent circuit.

15. Cascade the partial system transfer functions to generate the transfer function of the entire system.

16. Calculate the equivalent circuit performance parameters from static characteristics and pressures and flows at the operating bias points combined with effective dimensions of the components and their interconnections.

17. Substitute the calculated performance parameters in the transfer functions.

156   Detailed Systems Design Procedure

18. Calculate the analog system frequency response-estimate digital system transient response.
19. Compare the calculated response with the required performance specification.
20. Investigate individual component transfer functions and make the changes necessary to correct for deviations from desired performance.
21. Finalize preliminary design and calculate performance.
22. Generate a listing of factors important to achieving design goals (short small lines, symmetry, isolated supplies, adequate vents).

10.3 DESIGN CHECK LIST

A design check list is given below for the convenience of the system designer who is intimately familiar with the detailed design procedure.

1. Choose final control element (output).
2. Determine its input characteristics.
3. Choose suitable sensor (input).
4. Determine its output characteristics.
5. Block out system.
6. Choose fluidic devices.
7. Superimpose mating characteristics.
8. Explore matching problems.
9. Make operating points and ranges coincide.
10. Calculate transfer curves.
11. Investigate nonlinearities.
12. Develop equivalent circuits for components.
13. Prepare equivalent circuit for system.
14. Derive transfer functions of isolated networks.
15. Generate transfer function of system.
16. Calculate equivalent circuit parameters.
17. Substitute into transfer function.
18. Calculate system response.
19. Compare with specifications.
20. Make necessary changes.
21. Finalize design.
22. Generate list of critical factors.

Appendix A

Applicable Standards*

A.1 TERMINOLOGY

General

Fluidics. The general field of fluid devices or systems performing sensing, logic, amplification, and control functions employing primarily no-moving-part (flueric) devices.

Flueric. An adjective applied in some quarters to fluidic devices and systems performing sensing, logic, amplification, and control functions using no moving mechanical elements.

Elements. The general class of devices in their simplest form used to make up fluidic components and circuits; for example, fluidic restrictors and capacitors. These are the "least common denominators" of the fluidics technology.

Components. Fluidic devices that are interconnected with elements to form working circuits; for example, a proportional amplifier or an OR-NOR logic gate.

Analog. The general class of devices or circuits whose output is utilized as a continuous function of its input; for example, a proportional amplifier.

Digital. The general class of devices or circuits whose output is utilized as a discontinuous function of its input, for example, a bistable amplifier.

Active. The general class of devices that control power from a separate supply.

*From Ref. 1.

Passive. The general class of devices that operate on the signal power alone.

Bias (Quiescent) Point. The point on its static pressure-flow characteristics at which an element or component will be at equilibrium in a circuit with no signal applied.

Operating Range. The maximum range of signal over which an element or component is recommended for operation in a circuit. In an amplifier it is normally limited to the range where the gain can be considered linear.

Impedance. An effective restriction in the flow path.

Loading. Loading of a component is related to the output flow demanded from it in a circuit. For example, an amplifier is unloaded when it has an infinite impedance connected to its output port (blocked).

Amplifiers

Amplifier. An active fluidic component whose output signal is greater than its input signal.

Pressure Amplifier. A component designed specifically for amplifying pressure signals.

Flow Amplifier. A component designed specifically for amplifying flow signals.

Power Amplifier. A component designed specifically for amplifying power signals.

Vented versus Closed Amplifier. A vented amplifier utilizes auxiliary ports to establish a reference pressure in a particular region of the amplifier geometry; a closed amplifier has no communication with an independent reference. Terminology for the geometry is defined in Figures A.1 and A.2.

Jet-Interaction Amplifier. An amplifier that utilizes control jets to deflect a power jet and to modulate the output. Usually employed as an analog amplifier. Terminology for the geometry is defined in Figure A.1.

Wall-Attachment Amplifier. An amplifier that utilizes control of the *attachment* of a free jet to a wall (Coanda effect) to modulate the output. Usually employed as a digital amplifier. Terminology for the geometry is defined in Figure A.3.

Vortex Amplifier. An amplifier that utilizes the pressure drop across a controlled vortex for modulating the output. Terminology for the geometry is defined in Figure A.4.

Boundary-Layer-Control Amplifier. An amplifier that utilizes the control of the *separation* point of a power stream from a curved or plane surface to modulate the output. Terminology for the geometry is defined in Figure A.5.

Turbulence Amplifier. An amplifier that utilizes control of the laminar-to-turbulent transition of a

Terminology 159

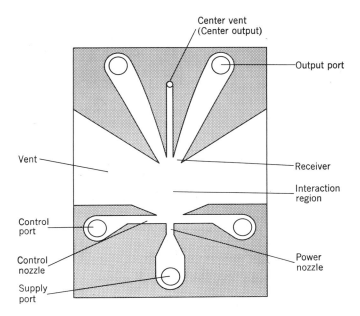

Figure A.1 Vented jet-interaction amplifier.

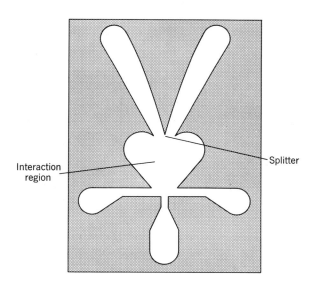

Figure A.2 Closed jet-interaction amplifier.

160 Applicable Standards

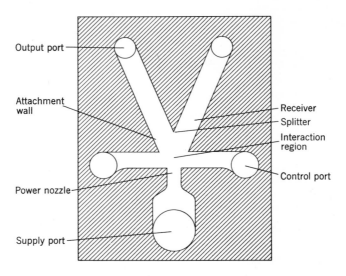

Figure A.3 Wall-attachment amplifier.

power jet to modulate the output. Terminology for the geometry is defined in Figure A.6.

Axisymmetric Focused-Jet Amplifier. An amplifier that utilizes control of the attachment of an annular jet to an axisymmetric flow separator (i.e., control of the focus of the jet) to modulate the output. Usually employed as a digital amplifier. Terminology is defined in Figure A.7.

Impact Modulator. An amplifier that utilizes the control of the in-

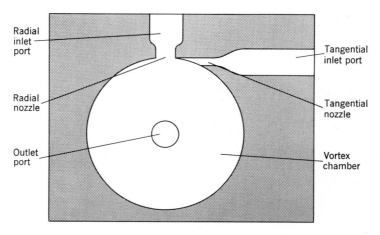

Figure A.4 Vortex amplifier.

Terminology 161

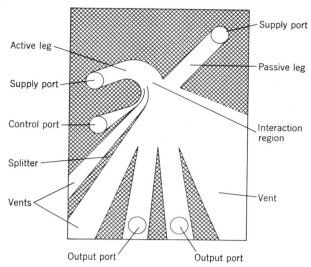

Figure A.5 Boundary-layer-control amplifier (vents optional).

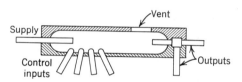

Figure A.6 Turbulence amplifier.

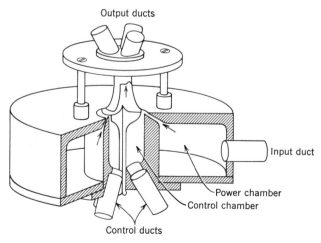

Figure A.7 Axisymmetric focused-jet amplifier.

tensity of two directly opposed, impacting power jets, thereby controlling the position of the impact plane to modulate the output. Terminology is defined in Figure A.8.

Sensors

Sensor. A component that, in general, senses variables and produces a signal in a medium compatible with fluidic devices, for example, a temperature or angular rate sensor.

Transducers

Transducer. A component that, in general, converts a signal from one medium to an equivalent signal in a second medium, one of which is compatible with fluidic devices.

Actuators

Actuator. A component that, in general, converts a fluidic signal into an equivalent mechanical output.

Displays

Display. A component that, in general, converts a fluidic signal into an equivalent visual output.

Logic Devices

Logic Device. The general category of digital fluidic components that perform logic functions; for example, AND, OR, NOR, and NAND. They can gate or inhibit signal transmission with the application, removal or other combinations of input signals.

Flip-Flop. A digital component or circuit with two stable states and sufficient hysteresis so that it has "memory." Its state is changed with an input pulse; a continuous input signal is not necessary for it to remain in a given state.

Circuit Elements

Impedance. A passive fluidic element that requires a pressure drop to establish a flow through it. Transfer function may have real and imaginary parts.

Resistor. Passive fluidic element that, because of viscous losses, produces a pressure drop as a function of the flow through it and has a transfer function of essentially real components (i.e., negligible phase shift) over the frequency range of interest.

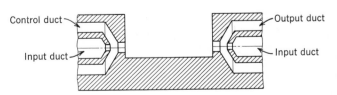

Figure A.8 Impact modulator.

Capacitor. A passive fluidic element that, because of fluid compressibility, produces a pressure that lags flow into it by essentially 90°.

Inductor. A passive fluidic element that, because of fluid inertance, has a pressure drop that leads flow by essentially 90°.

A.2 NOMENCLATURE AND UNITS

Basic Quantities

The quantities listed below are general; specific quantities should be identified by subscripts (e.g., P_{o2} would be pressure at port $O2$).

Quantity	Nomenclature	Units	
		Standard	SI
length	l	inch; in.	meter; m
force	F	pound; lb	newton; N
mass	m	lb-sec^2/in.	kilogram; kg
time	t	seconds; sec	seconds; sec
angle	—	degrees; °	radians; rad
frequency	f	hertz; Hz	hertz; Hz
area	A	in.^{2}	m^2
acceleration	a	in./sec^2	m/sec^2
temperature, static	T	degrees Rankine; °R	degrees Kelvin; °K
velocity, angular	ω	deg/sec;°/sec	rad/sec
acceleration, angular	α	deg/sec^2;°/sec^2	rad/sec^2
volume	V	in.^{3}	m^3
flow rate	Q	in.^{3}/sec (standard conditions); or scfm	m^3/sec (standard conditions)
velocity	v	in./sec	m/sec
pressure, general	P	lb/in.^{2}; or psi	N/m^2
pressure, absolute	P_a	psia	N/m^2
pressure, gage or drop	P_g	psig	N/m^2
fluid impedance	Z	lb sec/in.^{5}	Nsec/m^5
fluid resistance	R	lb sec/in.^{5}	Nsec/m^5
fluid capacitance	C	in.^{5}/lb	m^5/N
fluid inductance	L	lb sec^2/in.^{5}	Nsec/m^5

Applicable Standards

TABLE Contd

Quantity	Nomen-clature	Units Standard	SI
Laplace operator	s	1/sec	per sec
pressure gain, incremental	G_p		dimensionless
flow gain, incremental	G_f		dimensionless
power gain, incremental	G_w		dimensionless
signal-to-noise ratio	S/N		dimensionless

General Subscripts

control	c
output	o
supply	s
control bias	co
control differential	cd
output differential	od

A.3 GRAPHICAL SYMBOLOGY

It has been recognized that fluidic symbols were required to satisfy two basic needs. The first was the need of the system designer interested primarily in the *function* of the device. The second was the circuit designer interested primarily in the *operating principle* of the device.

Following is an integrated set of symbology that satisfies both requirements. Functions of the devices are defined by symbols enclosed within square envelopes. Operating principles of the devices are defined by symbols enclosed within round envelopes. The difference in envelopes is specifically intended to emphasize the difference in purpose of the symbols as shown below:

Functional symbol   Operating principle symbol

By definition the symbols are intended to show the following:

Functional symbol: Depicts a function that may be performed by a single fluidic element or by an interconnected circuit containing multiple elements.

Operating principle symbol: Depicts the fluid phenomena in the interaction region which is employed to perform the function, as well as the function of the fluidic element.

In the cases where no operating principle is indicated, it is implied that, at present, no *single* operating principle or interaction region is adequate to perform the function. In these cases a combination of operating principles or interaction regions is required to represent the function.

To further emphasize this point, consider the case of a Schmitt Trigger as follows:

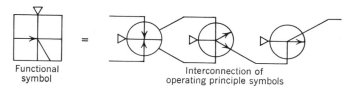

Functional symbol = Interconnection of operating principle symbols

General Conventions

Port Locations. The relative port locations for the symbols are patterned in the following manner:

All symbols may be oriented in 90° increments from the position shown.

Port Identification. Specific ports are identified by the following nomenclature:

Supply port — S
Control port — C
Output port — O

The nomenclature shown on the graphic symbols need not be used on schematic diagrams. It is primarily intended to correlate the function of each port with the truth table.

Active versus Passive. Supply ports can be either active or passive. An inverted triangle, $\triangledown$, denotes a supply source connected to the supply port (active device).

Active devices

166 Applicable Standards

Control Flow. An arrowhead on the control line inside the symbol envelope indicates continual flow is required to maintain state (no memory, no hysteresis):

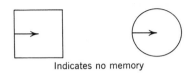
Indicates no memory

Interconnecting Lines. 1. Interconnecting fluid lines shall be shown with a dot at the point of interconnection:

2. Crossed fluid lines are to be shown without dots:

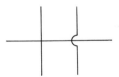

Start-up Condition. A small + on the output of a bistable device indicates initial or start-up flow condition.

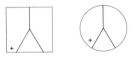

Analog Fluidic Devices

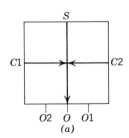
(a)

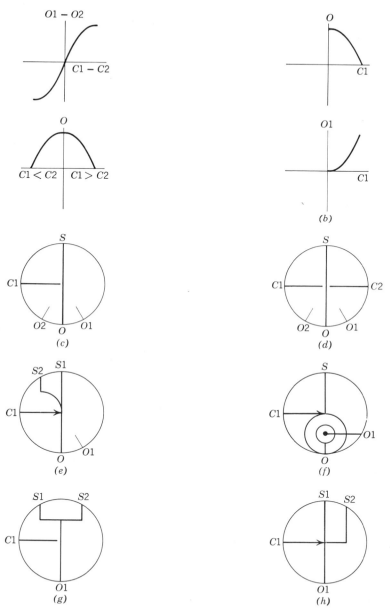

Figure A.9 Proportional amplifiers. (*a*) Functional symbol. (*b*) Functions. Operating principle symbols: (*c*) single input jet interaction; (*d*) differential jet interaction; (*e*) separation point control; (*f*) vortex; (*g*) transverse impact modulator; (*h*) direct impact modulator.

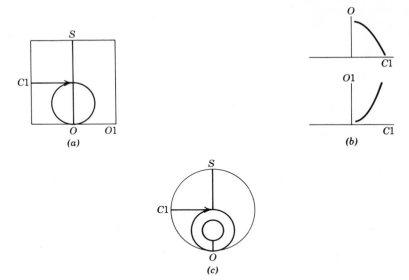

Figure A.10 Throttling valve. (*a*) Functional symbol. (*b*) Functions. (*c*) Operating principle symbol (vortex).

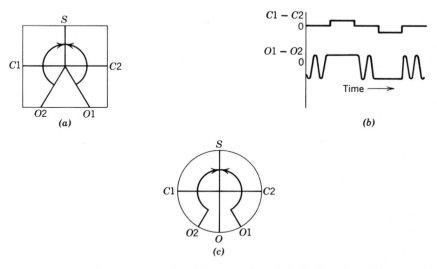

Figure A.11 Oscillator (sine wave). (*a*) Functional symbol. (*b*) Functions. (*c*) Operating principle symbol.

Graphical Symbology 169

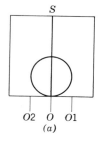
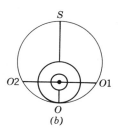

Figure A.12 Rate Sensor. (*a*) Functional symbol. (*b*) Operating principle symbol (vortex).

Bistable Digital Devices

	Truth Table		
C1	C2	O1	O2
1	0	1	0
0	0	1	0
0	1	0	1
0	0	0	1

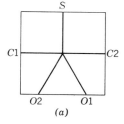
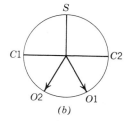

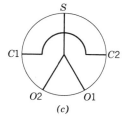
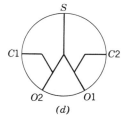

Figure A.13 Flip-flop. (*a*) Functional symbol. Operating principle symbols: (*b*) wall attachment; (*c*) edgetone; (*d*) induction.

170   Applicable Standards

Truth Table

C1	C2	O1	O2
1	0	1	0
0	1	0	1
0	0	Undefined	
1	1	Undefined	

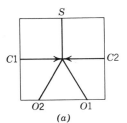
(a)

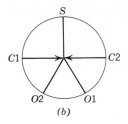
(b)

Figure A.14 Digital amplifier. (*a*) Functional symbol. (*b*) Operating principle symbol (jet interaction).

Truth Table

C	C1	C2	O1	O2
0	1	0	1	0
0	0	0	1	0
0	0	1	0	1
0	0	0	0	1
1	0	0	1	0
0	0	0	1	0
1	0	0	0	1
0	0	0	0	1

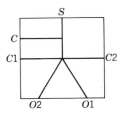

Figure A.15 Binary counter. Functional symbol.

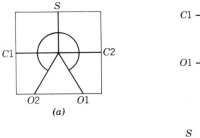

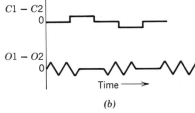

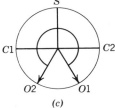
(c)

Figure A.16 Multivibrator. (*a*) Functional symbol. (*b*) Functions. (*c*) Operating principle symbol (wall attachment).

Monostable Digital Devices

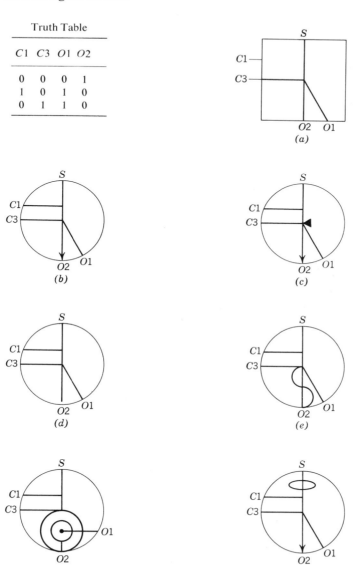

Figure A.17 OR-NOR. (*a*) Functional symbol. Operating principle symbols: (*b*) wall attachment; (*c*) wall attachment, internal fluid bias; (*d*) jet interaction, geometrical bias; (*e*) turbulence; (*f*) vortex; (*g*) focused jet.

Truth Table

C1	C3	O1	O2
0	0	0	1
1	0	0	1
0	1	0	1
1	1	1	0

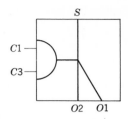

Figure A.18 AND-NAND. Functional symbol.

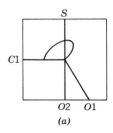
(a)

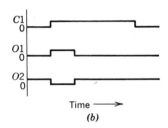
(b)

Figure A.19 One-shot. (a) Functional symbol; (b) Functions.

Truth Table

	O1	O2
C1 > C2	1	0
C1 < C2	0	1
C1 = C2	Undefined	

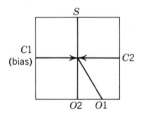

Figure A.20 Schmitt trigger. Functional symbol.

Truth Table

C1	C2	O1	O2
0	0	0	1
1	0	1	0
0	1	1	0
1	1	0	1

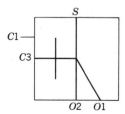

Figure A.21 Exclusive OR. Functional symbol.

Passive Digital Devices

Truth Table

C1	C3	O1
0	0	0
1	0	1
0	1	1
1	1	1

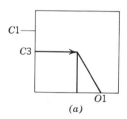
(a)

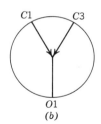
(b)

Figure A.22 OR. (a) Functional symbol. (b) Operating principle symbols (passive jet interaction).

Truth Table

C1	C3	O1	O2
1	0	1	0
0	1	1	0
1	1	0	1
0	0	0	0

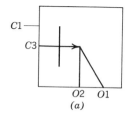
(a)

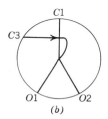
(b)

Figure A.23 Exclusive OR-AND. (a) Functional symbol. (b) Operating principle symbols (passive jet interaction).

Truth Table

C1	C2	O1	O2	O
1	0	1	0	0
0	1	0	1	0
1	1	0	0	1
0	0	0	0	0

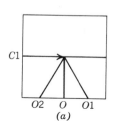
(a)

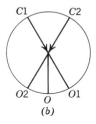
(b)

Figure A.24 AND-2/3AND. (a) Functional symbol. (b) Operating principle symbols (passive jet interaction).

Fluidic Impedances

Figure A.25 General resistance — fixed.

Figure A.26 General resistance — variable.

174 Applicable Standards

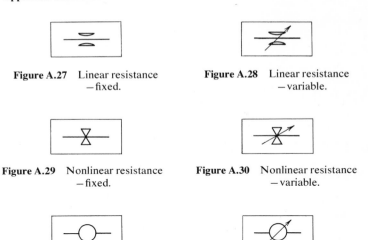

Figure A.27 Linear resistance — fixed.

Figure A.28 Linear resistance — variable.

Figure A.29 Nonlinear resistance — fixed.

Figure A.30 Nonlinear resistance — variable.

Figure A.31 Capacitance — fixed.

Figures A.32 Capacitance — variable.

Figure A.33 Inductance — fixed.

Figure A.34 Inductance — variable.

Figure A.35 Diode.

Appendix B

Illustrative Examples of V/Stol Control System Design

B.1 DESCRIPTION OF THE UH-1B YAW DAMPER SYSTEM

To illustrate the procedures described in this book, we have chosen to design an analog fluidic system to implement the UH-1B helicopter. The complete system is shown in Figure B.1 in block diagram form.

The fluidic portion of the system is redrawn in Figure B.2 with functional specifications noted.

Now to proceed with the design according to the methods given in the book; we will refer to Chapter 10 and follow the sequence outlined there.

B.2 REQUIRED INFORMATION

Performance Specifications for the System

1. Input range = ±25°/sec yaw rate (nominal).
2. Output range = ±0.38 in. deflection of servo.
3. Frequency response—characteristics of a highpass network with a time constant of 3.0 sec and less than 180° phase shift at 10 cps.
4. Linearity + 20%.
5. Threshold 0.10°/sec.

Characteristics of Available Supply, Signal Sources, and Driven Load

1. Supply—10 psi air at standard conditions.
2. Signal source—yaw rate of aircraft ±25°/sec with maximum of 40°/sec.

176 Examples of V/Stol Control System Design

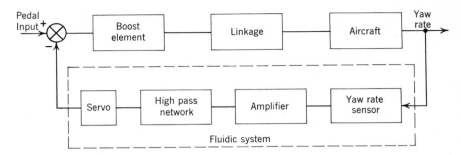

Figure B.1 Block diagram of UH-1B yaw damper system.

3. Driven load—mechanical linkage requiring ± 0.38 in. of deflection against a force matched by existing servo.

Static Characteristics of Available Components

1. Rate sensor—output characteristics shown in Figure B.3 (assumed balanced).
2. Amplifier A:
   Normalized output characteristics shown in Figure B.4 (assumed balanced).
   Normalized input characteristics shown in Figure B.5 (assumed balanced).
   Power nozzle characteristics shown in Figure B.6.
   Amplifier B:
   Normalized output characteristics shown in Figure B.7 (assumed balanced).
   Normalized input characteristics shown in Figure B.8 (assumed balanced).
   Power nozzle characteristics shown in Figure B.9.
3. Output servo actuator—input characteristics shown in Figure B.10 (essentially infinite input resistance).

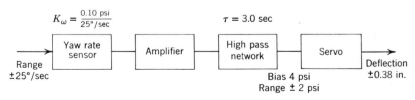

Figure B.2 Functional requirements of fluidic system.

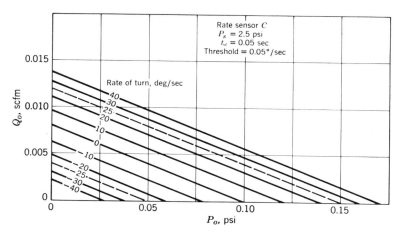

Figure B.3 Static output characteristics of rate sensor.

4. Laminar-flow restrictors — characteristics shown in Figure B.11 typical of all values of linear restrictors available.

Internal Dimensional Characteristics of Available Components

1. Rate sensor — time delay is given on the static characteristics, so vortex chamber dimensions are not necessary. Output tubes are 0.060 in. inside diameter and 2 in. long.

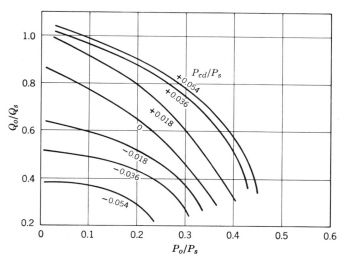

Figure B.4 Normalized output characteristics of amplifier A.

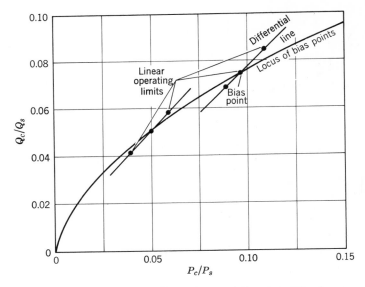

Figure B.5 Normalized input characteristics of amplifier A.

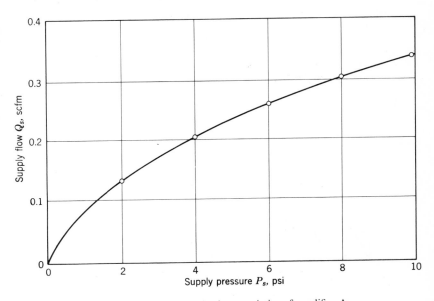

Figure B.6 Power nozzle characteristics of amplifier A.

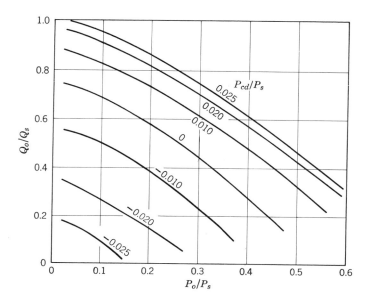

Figure B.7 Normalized output characteristics of amplifier B.

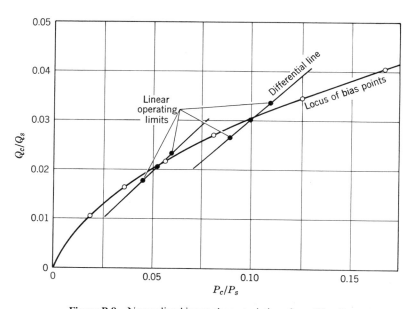

Figure B.8 Normalized input characteristics of amplifier B.

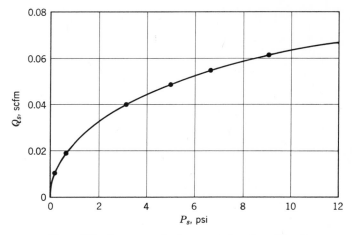

Figure B.9 Power nozzle characteristics of amplifier B.

2. Amplifier A:
   Dimensional sketch is shown in Figure B.12.
   Ferrules connected to input and output passages 0.5 in. long, 0.19 in. inside diameter.
   Amplifier B:
   Dimensional sketch is shown in Figure B.13.
   Ferrules connected to input and output passages 0.44 in. long, 0.13 in. inside diameter.
3. Output servo actuator — input volume approximately 0.001 in.³. Ferrules connected to input 0.44 in. long, 0.13 in. inside diameter.
4. Laminar restrictors — assume negligible delay time, volume, and length.

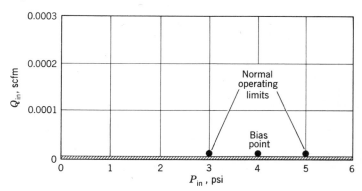

Figure B.10 Input characteristics of servo actuator.

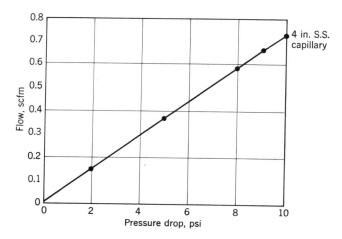

Figure B.11 Characteristics of typical laminar-flow restrictor.

B.3 STEP-BY-STEP DESIGN (according to Chapter 10)

1. Choose Final Control Element. The servo actuator for the fluidic UH-1B yaw damper is fixed by existing system considerations.

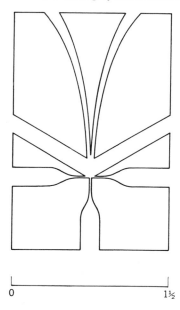

Figure B.12 Dimensional sketch of amplifier A (depth of channels 0.050).

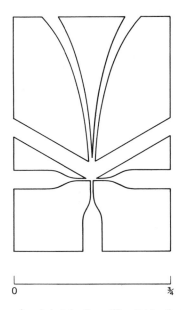

Figure B.13 Dimensional sketch of amplifier B (depth of channels 0.025).

2. Determine Its Input Characteristics. As given, the static input characteristics (Figure B.10) appear as an infinite resistance (no flow). Reflected load dynamics are negligible.

3. Choose a Suitable Sensor. The static characteristics of the available rate sensor (Figure B.3) show that it has the required sensitivity, linearity, range, and threshold.

4. Determine Its Output Characteristics. The static characteristics are given in Figure B.3. The time delay is given as measured by the supplier (50 msec). The output capacitance and inductance can be calculated from the dimensions of the output tubing.

5. Block Out the System between Sensor and Final Control Element. A preliminary schematic diagram of the fluidic system is shown in Figure B.14. Referring to Figure B.3 we see that the rate sensor has a maximum (blocked load) differential output of 0.10 psi for an input of 25°/sec. The performance specifications call for a servo output deflection of 0.38 in. for an input yaw rate of 25°/sec, which corresponds with a differential pressure at the servo input of 2 psi. In other words, we get 0.1 psi out of the rate sensor and we need 2 psi out of the fluidic amplifiers; therefore, with ideal matching, the minimum amplifier pressure gain must be 20.

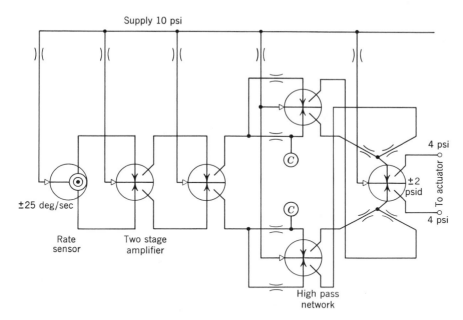

Figure B.14 Preliminary schematic diagram of fluidic yaw damper.

6. Tentatively Choose Fluidic Devices. We have tentatively selected two fluidic amplifiers, either of which may satisfy the requirements for pressure gain and matching. These are designated Amplifier A and Amplifier B, and their characteristics are described earlier. They are illustrated in Figures B.4 through Figure B.9.

Rate Sensor and First Stage Amplifier

7. and 8. Explore the Operating Bias Point Matching Problem. Examination of the rate sensor output characteristics (Figure B.3) for maximum pressure sensitivity reveals that we should operate near 0.10 psi. If we do so, the amplifier connected to it must have a bias pressure of 0.10 psi and *require no flow*. This is not realistic, so we will compromise for a bias point at 0.07 psi. Assuming a supply pressure of 10 times the input bias pressure (0.7 psi), we can calculate the input flow required by Amplifier A at an input bias pressure of 0.07 psi. That is, for

$$P_s = 0.7 \text{ psi}$$

$$Q_s = 0.60 \text{ scfm} \quad \text{(see Figure B.6)}$$

184 Examples of V/Stol Control System Design

and
$$\sqrt{Q_s} = 0.245$$
when
$$P_c/P_s = 0.10$$
then
$$Q_c/\sqrt{Q_s} = 0.077 \quad \text{(see Figure B.5)}$$
$$Q_c = 0.077 \times 0.245$$
$$Q_c = 0.0189 \text{ scfm}$$

Superimposing this point ($P_c = 0.07$ psi, $Q_c = 0.0189$ scfm) on the rate sensor characteristics (see Figure B.15) shows that the input flow required cannot be supplied by the rate sensor, and that the impedance match (relative slopes) is poor.

Drawing a rough square-law curve between the origin and the point just calculated ($P_c = 0.07$ psi, $Q_c = 0.018$ scfm) shows that the amplifier input bias will probably match the rate sensor output bias (zero signal line) around 0.01 psi. Again assuming that the amplifier would have a supply pressure 10 times the input bias,

$$P_s = 0.1 \text{ psi}$$
then
$$Q_s \cong 0.01 \text{ scfm} \quad \text{(see Figure B.6)}$$
$$\sqrt{Q_s} = 0.1$$

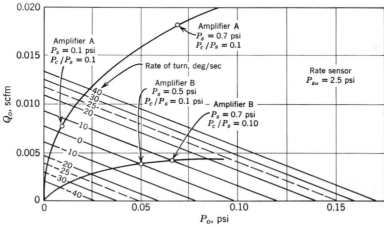

Figure B.15 Rate sensor output characteristics with first stage input bias characteristics superimposed.

when
$$P_c/P_s = 0.10$$
$$Q_c/\sqrt{Q_s} = 0.077 \quad \text{(see Figure B.5)}$$

then
$$Q_c = 0.077 \times 0.1$$
$$Q_c = 0.0077 \text{ scfm}$$

Plotting this point ($P_c = 0.01$ psi, $Q_c = 0.0077$ scfm) on the rate sensor characteristics (Figure B.15) shows that Amplifier A could be matched to the rate sensor using an amplifier supply pressure of 0.01 psi. However, with an amplifier supply pressure of 0.01 psi it would be impossible to get more than about 0.005 psi out of it. Therefore, we would be throwing away too much pressure sensitivity.

9. Make the operating bias points coincide. To rectify this situation, we will choose Amplifier B, which has smaller control nozzles and therefore higher input resistance (less input flow). Assuming a bias point of 0.07 psi, the supply will be

$$P_s = 0.7 \text{ psi}$$

then
$$Q_s = 0.019 \text{ scfm} \quad \text{(see Figure B.9)}$$
$$\sqrt{Q_s} = 0.138$$

when
$$P_c/P_s = 0.10$$
$$Q_c/\sqrt{Q_s} = 0.030 \quad \text{(see Figure B.8)}$$

then
$$Q_c = 0.030 \times 0.138$$
$$Q_c = 0.0042 \text{ scfm}$$

Plotting this point ($P_c = 0.07$ psi, $Q_c = 0.0042$ scfm) on the rate sensor output characteristics (Figure B.15) shows that the amplifier bias is too high. Drawing a rough square-law curve from the origin through this point will help to "zero-in" on the correct match. It appears at approximately 0.05 psi, where the amplifier input locus of bias points crosses the zero signal output line of the rate sensor. At this point

$$P_s = 0.5 \text{ psi} \quad \text{(10 times input bias)}$$

then

$$Q_s = 0.016 \text{ scfm} \quad \text{(see Figure B.9)}$$
$$\sqrt{Q_s} = 0.127$$

when

$$P_c/P_s = 0.10$$
$$Q_c/\sqrt{Q_s} = 0.030 \quad \text{(see Figure B.8)}$$

then

$$Q_c = 0.030 \times 1.127$$
$$Q_c = 0.0039 \text{ scfm}$$

Plotting this point ($P_c = 0.05$ psi, $Q_c = 0.0039$ scfm) on the rate sensor output characteristics (Figure B.15) shows that the operating bias points of the rate sensor and amplifier are now properly matched.

8. Explore the matching of preferred operating ranges. Figure B.16 is a replot of the rate sensor output characteristics with the input characteristics superimposed, including the differential input curve with preferred operating limits. Note that to stay within a reasonably linear range of operation of Amplifier B, the input swing must be limited to $\pm 1\% P_s$ ($\pm 2\% P_s$ control differential P_{cd}). However, as illustrated in Figure B.16 the rate sensor, operating between the $\pm 25°/\text{sec}$ lines, would produce a much larger change in amplifier control pressure than $\pm 1\% P_s$. Therefore, the circuit must be altered to prevent this.

9. Make the operating ranges coincide. Effectively, what we require here is to apply some circuit element to reduce the output swing of the rate

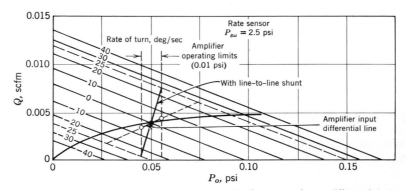

Figure B.16 Rate sensor output characteristics with first stage input differential characteristics superimposed.

sensor *without affecting the operating bias point*. The method for doing this is by means of a restrictor connected between the differential circuit lines (line-to-line). When there is zero input to the rate sensor, both output lines are biased at the same pressure (0.05 psi). Therefore, there is no flow through the restrictor connected between the output lines, and the circuit behaves as if it were not connected. When there is a signal out of the rate sensor, the output lines are unbalanced and there is flow through the restrictor. In this case it appears as a heavier load on the rate sensor, requiring more output flow for the same increase in output pressure. Graphically (as shown in Figure B.16) the effect is to rotate the *differential* load line around the operating bias point.

Now if we want to prevent the pressure swing from exceeding 0.005 psi ($\pm 1\% P_s$) when the rate sensor is driven between the extremes of $\pm 25°$/sec, it is a simple geometrical exercise to find the line that will pass through the allowable extremes and the operating bias point. The value of the fluid resistance that must be connected across the differential circuit can be calculated from the change in slope from the original differential input line (amplifier input nozzle only) to the final line that fits within the required limits.

In summary of the matching problem between the rate sensor and the first stage of amplification, we first explored the matching of operating bias points and found that Amplifier B could be properly matched by using a supply pressure of 0.5 psi. Second, we explored the matching of operating ranges and found that Amplifier B could be properly matched by adding a shunt restrictor between the rate sensor output lines. The value of the resistance required is approximately 0.4 times the value of the input nozzle resistance of Amplifier B (draws 2.5 times the flow of the nozzle for the same pressure).

First and Second Stage Amplifiers

7. Explore the operating bias point matching problem. Figure B.17 shows the output characteristics of the first stage amplifier. Again, a bias point pressure is assumed, the second stage supply pressure is made 10 times higher, and a corresponding control bias flow is calculated. If the resulting point does not fall on or near the zero signal line for the first stage output, a more realistic bias point pressure is assumed and the process is repeated. Thus we find a second stage amplifier supply pressure that will make it match with the first stage at the point where the second stage input bias is at 10% of the second stage supply pressure. This is illustrated in Figure B.17 for a second stage supply pressure of 1.75 psi.

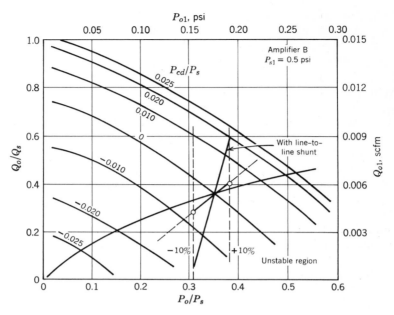

Figure B.17 First stage amplifier output characteristics with second stage input characteristics superimposed.

8. Explore the matching of preferred operating ranges. If we limit the allowable input signal swing of the second stage to $\pm 0.01 P_s$ (± 0.0175 psi), the first stage will overdrive the second stage (see Figure B.17). Therefore, it is necessary to reduce the first stage output swing by loading it with a restrictor connected across the differential circuit (line-to-line). Using a linear restrictor, the resulting load will be a line passing through the operating bias point and the intersections of the allowable extremes of first stage output and second stage input. The result is a differential load line, as shown in Figure B.17.

Note two important points. First, the extremes of output pressure are not equal with respect to the operating bias point, because the output characteristic curves are not evenly spaced. Second and most important, the situation illustrated is *not* a practical one because the lower end of the resulting load line passes through the unstable region of the output characteristics. Therefore, we must find a new operating bias point at a higher flow so that the amplifier will not have to operate in the unstable region.

9. Make the operating points coincide and the operating ranges compatible. By inspection of Figure B.17 it is clear that we must place the operating bias point at a higher flow level. This means that the slope of the locus of bias points must be raised; in other words, the resistance connected at the output of the first stage must be decreased. The obvious way to do this is by connecting a shunt restrictor to return (atmosphere). With a linear restrictor, the effect is to add a flow directly proportional to the pressure to the locus of bias points.

With reference to Figure B.18 we can estimate that if the output bias were about 0.14 psi, there would be room to get the differential load line in without running into the unstable operating region of the first stage amplifier. Then the second stage amplifier bias would be

$$P_c = 0.14 \text{ psi}$$

Then, if
$$P_s = 10 P_c$$
$$P_s = 1.4 \text{ psi}$$
$$Q_s = 0.027 \text{ cfm} \quad \text{(see Figure B.9)}$$
$$\sqrt{Q_s} = 0.164$$

when
$$P_c/P_s = 0.1$$
$$Q_c/\sqrt{Q_s} = 0.03 \quad \text{(see Figure B.8)}$$

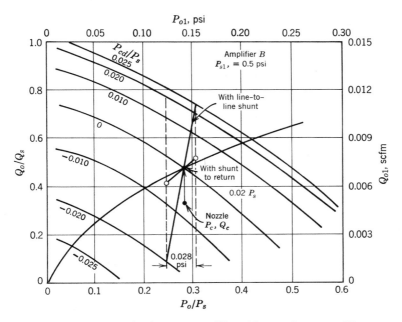

Figure B.18 Matching first stage amplifier with second stage amplifier.

then

$$Q_c = 0.03 \times 0.164$$
$$Q_c = 0.0049 \text{ scfm}$$

Again, with reference to Figure B.18 we must add approximately 0.0021 scfm to the flow out of the amplifier at 0.14 psi with the shunt restrictor to return. This places the operating bias point at $P = 0.14$ psi, and $Q = 0.007$ scfm. Now the differential control line with limits can be superimposed to show again that the first stage amplifier will overdrive the second stage amplifier. Then the differential control line must be rotated around the operating bias point by connecting a shunt restrictor line-to-line at the output of the first stage amplifier. The value of this restrictor is approximately 0.2 times the parallel resistance of control nozzle and shunt restrictor to return.

At this point we should stop to check if two stages of amplification will be enough to provide the proper overall gain. The rate sensor sensitivity loaded with the first stage amplifier is (see Figure B.16)

$$G_\omega = \frac{P_{od\omega}}{\omega} = \frac{0.010 \text{ psid}}{25°/\text{sec}}$$

The pressure gain of the first stage is

$$G_{p1} = \frac{P_{od1}}{P_{od\omega}} = \frac{0.028}{0.010} = 2.8$$

If we assume that the second stage will have the same gain as the first stage,

$$G_{p2} = 2.8$$

Then

$$\frac{P_{od2}}{\omega} = G_\omega \times G_{p1} \times G_{p2}$$

$$\frac{P_{od2}}{\omega} = \frac{0.010 \text{ psid}}{25°/\text{sec}} \times 2.8 \times 2.8$$

$$\frac{P_{od2}}{\omega} = \frac{0.078 \text{ psid}}{25°/\text{sec}}$$

We need (see Figure B.2) 2 psid/25°/sec; therefore, at least one more stage of amplification will be necessary.

Second and Third Stage Amplifiers

Since the procedure for the second and third stage amplifiers is the same as that given for the first and second, the exploratory work (Steps 7 and 8) will be omitted here.

9. Make the operating points coincide and the operating ranges compatible. With reference to Figure B.19 the matching problem, considered on a normalized basis, is almost identical with the situation encountered in the previous stage. Therefore, we can assume that a shunt restrictor to return will be required and that the bias will be at the same *normalized* point on the characteristics.

The supply pressure to stage 2 is 1.4 psi. Therefore, the actual output bias pressure will be 0.39 psi. Then for stage 3,

$$P_c = 0.39 \text{ psi}$$

if

$$P_s = 10 P_c$$
$$P_s = 3.9 \text{ psi}$$
$$Q_s = 0.043 \text{ scfm} \quad \text{(see Figure B.9)}$$
$$\sqrt{Q_s} = 0.206$$

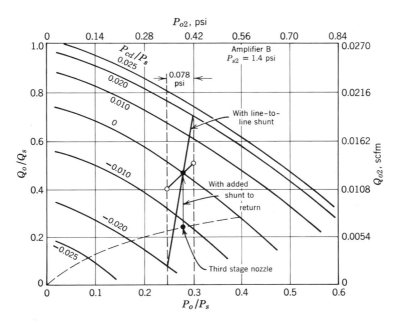

Figure B.19 Second stage amplifier with third stage amplifier.

when

$$P_c/P_s = 0.10$$
$$Q_c/\sqrt{Q_s} = 0.03 \quad \text{(see Figure B.8)}$$

then

$$Q_c = 0.03 \times 0.206$$
$$Q_c = 0.0062$$

To raise the bias flow to the required level of 0.0124 scfm, a restrictor must be shunted from the output of the stage 2 amplifier to return. The value of the restrictor is equal to the resistance of the stage 3 input nozzle at that pressure.

Again, to match the operating *ranges* it is necessary to connect a shunt restrictor line-to-line at the amplifier output. The value of the restrictor is equal to 0.2 times the resistance of the stage 3 input nozzle differential curve. The gain of the second stage is

$$G_{p2} = \frac{P_{od2}}{P_{od1}} = \frac{0.078 \text{ psi}}{0.028} = 2.8$$

Third and Fourth Stage Amplifiers

With reference to the circuit diagram shown in Figure B.14 the third stage must feed two parallel amplifiers in the high pass network. Another feature of this interface is that we prefer to have a restrictor in *series* with the fourth stage control nozzles. Since the network time constant is a function of the series resistance and an added volume capacitance, the larger we can make this resistance, the smaller will be the volume required.

7. *Explore the operating bias point matching problem*

8. *Explore the matching of operating ranges.* Referring to Figure B.20 the output characteristics of stage 3 are of the same form as stages 2 and 1. Also, we want to maintain the same $\pm 10\% P_c$ signal change around the operating bias point. Therefore, the *final* matching picture must be similar to Figure B.19, that is, the third stage output bias pressure should be about $P_o/P_s = 0.28$ (1.08 psi).

Suppose we want a series resistance of about 1.5 times the stage 4 nozzle resistance at the bias point. This means that only 2/5 of the pressure out of stage 3 appears at the input nozzle of stage 4 (2/5 × 1.08 psi = 0.43 psi). If we continue the practice of making the supply pressure 10 times the control bias pressure, the supply pressure of stage 4 would then

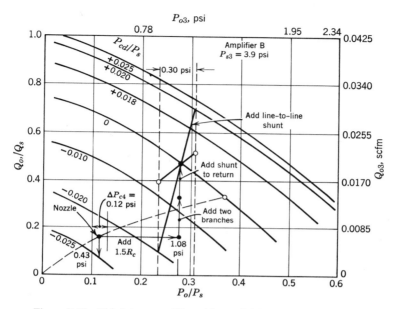

Figure B.20 Third stage amplifier with parallel fourth stage amplifiers.

be 4.3 psi. However, to illustrate a different situation (and to get more pressure gain), we will arbitrarily make the supply pressure of stage 4 = 6.0 psi. Then

$$P_s = 6.0 \text{ psi} \qquad P_c = 0.43 \text{ psi}$$
$$Q_s = 0.05 \text{ scfm} \qquad \text{(see Figure B.9)}$$
$$\sqrt{Q_s} = 0.223$$

Now

$$P_c/P_s = 0.43/6.0 = 0.072$$

and

$$Q_c/\sqrt{Q_s} = 0.025 \qquad \text{(see Figure B.8)}$$

then

$$Q_c = 0.025 \times 0.223$$
$$Q_c = 0.0056 \text{ scfm}$$

9. Make the operating bias points coincide and the operating ranges compatible. This point ($P_{c4} = 0.43$ psi, $Q_{c4} = 0.0056$ scfm) is plotted on Figure B.2. When 1.5 times the resistance represented by this point is added in series, the overall pressure is increased 1.5 times for the same flow (as illustrated) to $5/2 \times 0.43 = 1.08$ psi. This represents the total resistance seen looking into one branch of the input to one stage 4 amplifier.

As shown in Figure B.14 there are two branches to input nozzles of a single stage 4 connected to one output port of the stage 3 amplifier. Therefore, twice the flow calculated for one branch must be supplied by one output, and the point illustrated in Figure B.20 is actually at $P_{o3} = 1.08$ psi and $Q_{o3} = 0.0112$ scfm. But this is not yet enough flow to satisfy the requirements for the operating bias point ($P_{o3} = 1.08$ psi, $Q_{o3} = 0.0198$ scfm) we have chosen. Therefore, a shunt restrictor must be connected between each stage 3 output port and the return of a value that will carry the difference in flow (0.0086 scfm) with a pressure drop of 1.08 psi.

To limit the input swing to stage 4 to $\pm 0.01\, P_s$ (0.12 psi), the output from stage 3 must be limited to the same ratio of the bias point, $0.12/0.43 = 0.28$. Then, as shown in Figure B.20 the swing must be limited to $0.28 \times 1.08 = 0.30$ psi. Again this requires a line-to-line shunt restrictor of a value which will draw an additional flow of 0.0199 scfm at a pressure differential of 0.30 psi.

Fifth Stage Amplifier with Servo Load

Before matching the fourth stage to the fifth stage, we must first define the input requirements of the fifth stage. These are determined by the output required to drive the servo actuator. The process of matching stage 5 to the servo actuator is illustrated in Figure B.21.

7. Explore the operating bias point matching problem. The servo requires an input bias of 4.0 psi and has an infinite input resistance. By inspection of this situation, as shown in Figure B.21, it is clear that some measures must be taken to load the amplifier so that it is not required to operate in the unstable region.

8. Explore the matching of preferred operating ranges. It is required that the stage 5 amplifier swing from 5.0 psi to 3.0 psi (4.0 ± 1.0 psi). Inspection of Figure B.21 makes it clear that, with a high impedance load, the required output swing can be achieved with less than the maximum allowable input swing.

9. Make the operating bias points coincide and the operating ranges compatible. The first consideration is to get the operating bias point into a stable region of amplifier operation. This can be accomplished (as shown in Figure B.21) with a shunt restrictor to return, which raises the slope of the effective load line. A stable point is where the zero signal line crosses the $P_o/P_s = 0.4$ grid line; and since we need an output bias of 4.0 psi, a

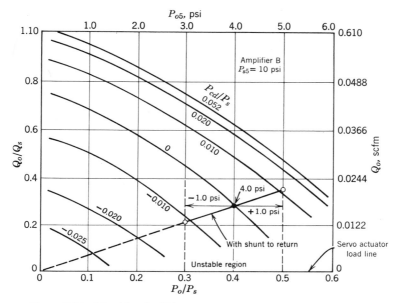

Figure B.21 Matching the fifth stage amplifier with the servo actuator.

convenient supply pressure is 10.0 psi. The shunt restrictors to return, connected at each output port, must draw 0.0165 scfm at 4.0 psi.

Drawing a linear load line through the operating bias point shows the range of output swing that can be achieved with a particular input swing. It is clear that the required output swing can be achieved with a control differential pressure of approximately $0.012\ P_s$ (0.12 psi). This is well within the linear range of Amplifier B.

Fourth and Fifth Stage Amplifiers

7. Explore the operating bias point matching problem. Figure B.22 shows the output characteristics of a fourth stage amplifier. To these we must match the input to stage 5, which, because it has a supply pressure of 10 psi, should be between 0.5 psi and 1.0 psi $(5 - 10\% P_s)$.

Three other features are important to note; first, two stage 4 amplifiers feed a single stage 5 amplifier; second, we need some series resistance to isolate one fourth stage amplifier from the other and to sum their outputs effectively; and finally, the operating bias point must be raised to a point where the extremes of operation will not be in an unstable region.

8. Explore the matching of preferred operating ranges. The input of stage 4 is driven ± 0.12 psi (see Figure B.20) or to $P_{cd}/P_s = 0.02$. The input to

196   Examples of V/Stol Control System Design

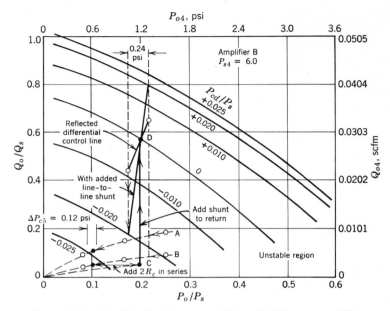

Figure B.22   Matching fourth stage amplifiers with fifth stage amplifier.

stage 5 cannot be driven more than 0.12 psi. Therefore, with the pressure scale shown in Figure B.22 it is clear that a line-to-line shunt restrictor will be required.

9. Make the operating bias points coincide and the operating ranges compatible. The fifth stage amplifier has the following input bias requirements:

$$P_s = 10 \text{ psi}$$
$$Q_s = 0.061 \text{ scfm} \quad \text{(see Figure B.9)}$$
$$\sqrt{Q_s} = 0.246$$

If we choose

$$P_c/P_s = 0.06 \text{ (0.6 psi)}$$
$$Q_c/\sqrt{Q_s} = 0.022 \quad \text{(see Figure B.8)}$$

then

$$Q_c = 0.022 \times 0.246$$
$$Q_c = 0.0054$$

This point ($P_c = 0.6$ psi, $Q_c = 0.0054$ scfm) is shown as point A on Figure B.22. Since two stage 4 amplifiers are supplying the stage 5 input requirements, each stage 4 amplifier need only supply half the flow. Therefore, it would see a point represented as point B in Figure B.22, which is equivalent to a resistance twice the nozzle resistance of amplifier B ($2R_{c5}$). For isolation, we insert a series restrictor equal to twice the nozzle resistance ($2R_{c5}$), which doubles the pressure required to supply the necessary flow and makes the effective load line on stage 4 pass through point C. This is, of course, in the unstable region, so we must add a shunt restrictor to return, with a value of resistance of about $\frac{1}{5}R_{c5}$. This matches the operating bias points of stages 4 and 5 (point D).

The operating range at the input of stage 5 must be limited to 0.12 psi. Since twice this pressure is required at the output of stage 4 (because of the series resistance), there it must be limited to 0.24 psi. As shown in Figure B.22 the reflected differential control line would not be operated within these limits; therefore, a line-to-line shunt restrictor is required with a resistance of approximately $\frac{1}{2}R_{c5}$. Then the operating range of stage 4 will be properly matched with the operating range of stage 5.

Complete Fluidic System

The schematic diagram of the complete matched fluidic yaw damper system is shown in Figure B.23.

10. Calculate the transfer curve of the system. In the process of matching operating ranges, we have automatically generated the proper overall transfer curve. That is, we have deliberately designed the system so that we get a differential pressure to the servo of 2 psid when the rate of turn is 25°/sec.

The actual overall transfer curve calculated point-by-point from the curves in Figures B.16 through B.22 is shown in Figure B.24.

11. Investigate linearity and inefficient use of operating ranges. With reference to the transfer curve of Figure B.24 the requirements for linearity (+20%) have been surpassed.

With respect to efficient use of operating ranges, a review of Figures B.16 through B.22 shows that in every case but one, the maximum allowable operating range has been used. In stage 5 we have limited it to $\pm 0.012 P_s$ rather than $\pm 0.020 P_s$ to avoid overdriving the servo actuator. However, this represents a source of additional linear gain, if it is needed in the final adjustment of performance of the yaw damper system.

12. Select the appropriate equivalent electrical circuit for each component of the system.
   a. Rate Sensor: The equivalent circuit for the rate sensor is identical to that shown in Figure 8.7 with external resistor added.
   b. Amplifiers: The equivalent circuit for the amplifier is identical

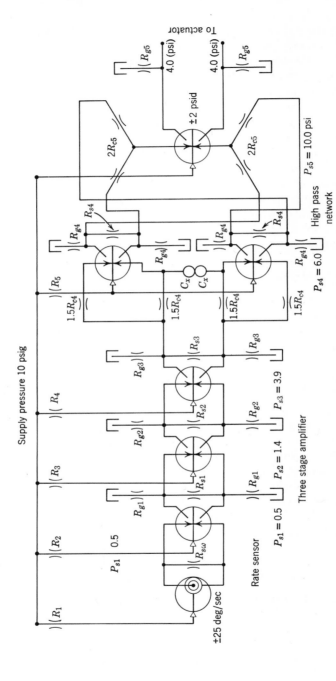

Figure B.23 Schematic diagram of complete matched fluidic yaw damper.

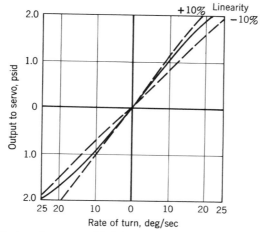

Figure B.24 Static transfer curve of fluidic yaw damper without high pass characteristic.

to the circuit for the vented jet-interaction amplifier shown in Figure 8.3. External resistors and capacitors are added to the appropriate stages.

c. *Actuator:* The equivalent electrical circuit for the actuator is represented simply by a volume capacitance. The input resistance is infinite, and the inductance will be considered negligible.

13. Prepare an equivalent electrical circuit of the entire coupled system. The total equivalent circuit, including all matching restrictors and added volume capacitors, is shown in Figure B.25. Inductances have been neglected for simplicity; they are later shown to be negligible.

14. Derive the transfer function of each portion of the equivalent circuit. The transfer function of each portion of the circuit between equivalent generators can be developed independently. We will illustrate the process by considering the network between the third stage and the two parallel fourth stages (see Figure B.26).

The network can be consolidated as shown in Figure B.27. Then by simple circuit analysis,

$$\frac{P_x}{P} = \frac{\dfrac{Z_p R_{g3}}{sC_{o3}}}{\dfrac{Z_p R_{g3} + \dfrac{R_{g3}}{sC_{o3}} + \dfrac{Z_p}{sC_{o3}}}{R_{o3} + \dfrac{\dfrac{Z_p R_{g3}}{sC_{o3}}}{Z_p R_{g3} + \dfrac{R_{g3}}{sC_{o3}} + \dfrac{Z_p}{sC_{o3}}}}}$$

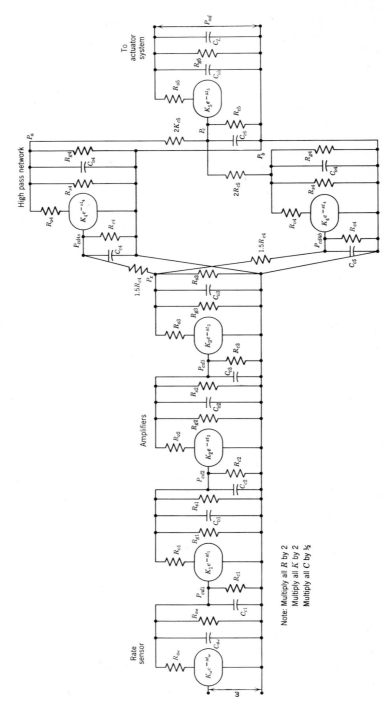

Figure B.25 Equivalent electric circuit for the complete fluidic yaw damper.

Step-by-Step Design 201

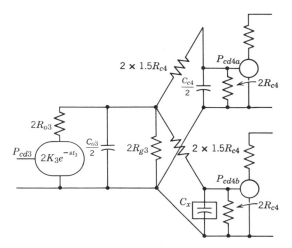

Figure B.26 Detail of network between stage 3 and stage 4.

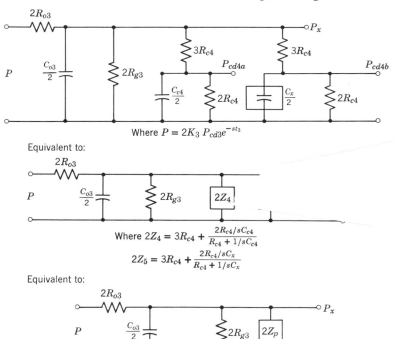

Figure B.27 Consolidation of the network between stage 3 and stage 4.

which reduces to

$$\frac{P_x}{P} = \frac{Z_p R_{g3}}{R_{o3}R_{g3} + R_{o3}Z_p + Z_p R_{g3}} \left(1 + \frac{sC_{o3}R_{o3}Z_p R_{g3}}{R_{o3}R_{g3} + R_{o3}Z_p + Z_p R_{g3}}\right)^{-1}$$

but

$$P = 2P_{cd3}K_3 e^{-st_3}$$

then

$$\frac{P_x}{P_{cd3}} = \frac{2K_3 Z_p R_{g3} e^{-st_3}}{R_{o3}R_{g3} + R_{o3}Z_p + Z_p R_{g3}} \left(1 + sC_{o3}\frac{R_{o3}Z_p R_{g3}}{R_{o3}R_{g3} + R_{o3}Z_p + Z_p R_{g3}}\right)^{-1}$$

Referring again to Figure B.27, we have

$$\frac{P_{cd4a}}{P_x} = \frac{\dfrac{R_{c4}}{sC_{c4}}}{R_{c4} + \dfrac{1}{sC_{c4}}}}{1.5R_{c4} + \dfrac{\dfrac{R_{c4}}{sC_{c4}}}{R_{c4} + \dfrac{1}{sC_{c4}}}}$$

which reduces to

$$\frac{P_{cd4a}}{P_x} = \frac{1}{2.5}(1 + 0.6sC_{c4}R_{c4})^{-1}$$

where C_{c4} is the normal circuit volume capacitance.

Similarly, we have the network leading from P_x to P_{cd4b}, which is analyzed in the same fashion to give

$$\frac{P_{cd4b}}{P_x} = \frac{1}{2.5}(1 + 0.6sC_x R_{c4})^{-1}$$

where C_x is the total volume capacitance in the circuit, some of which is added volume to generate the highpass filter network time constant (3.0 sec).

Each isolated portion of the circuit diagram in Figure B.25 is analyzed in the same way. The isolated transfer functions are shown in block diagram form in Figure B.28. The transfer functions corresponding to the individual blocks are listed in Figure B.29.

15. Cascade the partial system transfer functions to generate the transfer function for the entire system. With reference to the block diagram in Fig-

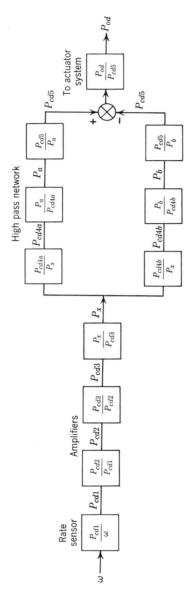

Figure B.28 Transfer function block diagram for the complete fluidic yaw damper.

Examples of V/Stol Control System Design

Rate Sensor

$$\frac{P_{cd1}}{\omega} = \frac{2K_\omega R_{c1} R_{s\omega} e^{-st_\omega}}{R_{c1}R_{s\omega}+R_{o\omega}R_{c1}+R_{o\omega}R_{s\omega}} \left(1+s(C_{o\omega}+C_{c1})\frac{R_{c1}R_{o\omega}R_{s\omega}}{R_{c1}R_{o\omega}+R_{o\omega}R_{s\omega}+R_{c1}R_{s\omega}}\right)^{-1}$$

Stage 1

$$\frac{P_{cd2}}{P_{cd1}} = \frac{2K_1 R_{c2} R_{s1} e^{-st_1}}{R_{c2}R_{o1}+R_{o1}R_{s1}+R_{c2}R_{s1}} \left(1+s(C_{o1}+C_{c2})\frac{R_{c2}R_{o1}R_{s1}}{R_{c2}R_{o1}+R_{o1}R_{s1}+R_{c2}R_{s1}}\right)^{-1}$$

Stage 2

$$\frac{P_{cd3}}{P_{cd2}} = \frac{2K_2 R_{c3} R_{s2} e^{-st_2}}{R_{c3}R_{o2}+R_{o2}R_{s2}+R_{c3}R_{s2}} \left(1+s(C_{o2}+C_{c3})\frac{R_{c3}R_{o2}R_{s2}}{R_{c3}R_{o2}+R_{o2}R_{s2}+R_{c3}R_{s2}}\right)^{-1}$$

Stage 3

$$\frac{P_x}{P_{cd3}} = \frac{2K_3 Z_p R_{g3} e^{-st_3}}{R_{o3}R_{g3}+R_{o3}Z_p+Z_o R_{g3}} \left(1+sC_{o3}\frac{R_{o3}Z_p R_{g3}}{R_{o3}R_{g3}+R_{o3}Z_p+Z_p R_{g3}}\right)^{-1}$$

where

$$Z_p = \frac{Z_4 Z_5}{Z_4 + Z_5}$$

$$Z_4 = 1.5R_{c4} + \frac{\dfrac{R_{c4}}{sC_{c4}}}{R_{c4}+\dfrac{1}{sC_{c4}}}$$

$$Z_5 = 1.5R_{c4} + \frac{\dfrac{R_{c4}}{sC_x}}{R_{c4}+\dfrac{1}{sC_x}}$$

also

$$\frac{P_{cd4a}}{P_x} = \frac{1}{2.5}(1+0.6sC_{c4}R_{c4})^{-1}$$

and

$$\frac{P_{cd4b}}{P_x} = \frac{1}{2.5}(1+0.6sC_x R_{c4})^{-1}$$

where C_x is determined by external volume added.

Stage 4

$$\frac{P_a}{P_{cd4a}} = \frac{2K_4 Z_c R_{g4} e^{-st_4}}{R_{o4}R_{g4}+Z_c R_{o4}+Z_c R_{g4}} \left(1+sC_{o4}\frac{Z_c R_{g4}R_{o4}}{R_{o4}R_{g4}+Z_c R_{o4}+Z_c R_{g4}}\right)^{-1}$$

where

$$Z_c = 2R_{c5} + Z_b$$

$$Z_b = \frac{\dfrac{Z_a R_{c5}}{sC_{c5}}}{Z_a R_{c5} + \dfrac{R_{c5}}{sC_{c5}} + \dfrac{Z_a}{sC_{c5}}}$$

$$Z_a = 2R_{c5} + \frac{\dfrac{R_{o4}}{sC_{o4}}}{R_{o4}+\dfrac{1}{sC_{o4}}}$$

and

$$\frac{P_{cd5}}{P_a} = \frac{Z_a}{2R_{c5}+3Z_a}\left(1+sC_{c5}\frac{2R_{c5}Z_a}{2R_{c5}+3Z_a}\right)^{-1}$$

also

$$\frac{P_b}{P_{cd4b}} = \frac{2K_4 Z_c R_{g4} e^{-st_d}}{R_{o4}R_{g4}+Z_c R_{o4}+Z_c R_{g4}}\left(1+sC_{o4}\frac{Z_c R_{g4} R_{o4}}{R_{o4}R_{g4}+Z_c R_{o4}+Z_c R_{g4}}\right)^{-1}$$

and

$$\frac{P_{cd5}}{P_b} = \frac{Z_a}{2R_{c5}+3Z_a}\left(1+sC_{c5}\frac{2R_{c5}Z_a}{2R_{c5}+3Z_a}\right)^{-1}$$

Stage 5

$$\frac{P_{od}}{P_{cd5}} = \frac{2K_5 R_{L5}}{R_{L5}+R_{o5}}\left(1+s(C_{o5}+C_L)\frac{R_{o5}R_{g5}}{R_{g5}+R_{o5}}\right)^{-1}$$

Figure B.29 List of transfer functions for the complete fluidic yaw damper system.

ure B.28, the overall transfer function for the fluidic yaw damper system can be developed as follows:

$$\frac{P_{od}}{\omega} = \frac{P_{cd1}}{\omega}\cdot\frac{P_{cd2}}{P_{cd1}}\cdot\frac{P_{cd3}}{P_{cd2}}\cdot\frac{P_x}{P_{cd3}}\left[\frac{P_{cd4a}}{P_x}\cdot\frac{P_a}{P_{cd4a}}\cdot\frac{P_{cd5}}{P_a}-\frac{P_{cd4b}}{P_x}\cdot\frac{P_b}{P_{cd4b}}\cdot\frac{P_{cd5}}{P_b}\right]\frac{P_{od}}{P_{cd5}}$$

Because of the complexity of the result, we will not substitute the individual transfer functions into this expression here, but the process is a straightforward one. It will also be more convenient to evaluate the frequency response of each one separately, then cascade the numerical results.

16. Calculate the equivalent circuit parameters. The values of the elements of the equivalent circuits and the parameters in the transfer functions are calculated according to the procedures described in Chapter 9. We will choose a few examples for the yaw damper system to illustrate the approach in detail. Consider first the circuit associated with stages 2 and 3, assuming a physical layout as shown in Figure B.30.

a. Pressure amplification factor K_2. The pressure amplification factor is defined as

$$K_p = \frac{\Delta P_o}{\Delta P_{cd}}\bigg|_{Q_o \text{ constant}}$$

and is calculated at the operating bias point. With reference to Figure B.19 we take the horizontal distance between the $P_{cd}/P_s = \pm 0.01$ lines as the change in output:

$$\Delta P_o = 0.56 - 0.20 \text{ psi}$$
$$\Delta P_o = 0.36 \text{ psi}$$

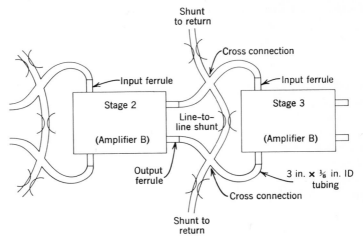

Figure B.30 Physical layout of circuit associated with stage 2.

The corresponding change in control differential is

$$\Delta P_{cd} = 0.02\, P_s$$

and the supply pressure for this stage is 1.4 psi

$$\Delta P_{cd} = 0.02 \times 1.4 = 0.028$$

Then

$$\frac{\Delta P_o}{\Delta P_{cd}} = \frac{0.36}{0.028}$$

$$\therefore K_2 = 12.8$$

b. *Output resistance* R_{o2}. The output resistance is defined as

$$R_o = \left.\frac{\Delta P_o}{\Delta Q_o}\right|_{P_o \text{ constant}}$$

and is calculated at the operating bias point. With reference to Figure B.19 the output resistance is the inverse of the slope of the zero signal line at the operating point, or

$$R_{o2} = \frac{32 \times 0.014 \text{ psi}}{23 \times 0.00054 \text{ scfm}} \times \frac{60}{1728}$$

$$\therefore R_{o2} = 1.26 \frac{\text{lb sec}}{\text{in.}^5}$$

c. *Output capacitance C_{o2}*. The output capacitance is defined as

$$C_o = \frac{V}{P_{abs}}$$

calculated at the operating bias point. With reference to Figure B.19 the operating pressure of 0.39 psi gage or 15.09 psi absolute.

The volume V charged to the amplifier output is the volume of the output aperture plus the volume of the connecting ferrule. With reference to Figure B.13 the volume of the output aperture is estimated from an integration of the plan area (by counting small dimensional blocks) times the depth of channel. In this case, the volume of the output aperture is 0.00057 in.^{3}. Therefore,

$$C_{o2} = \frac{0.00057 + 0.0053}{15.09} = \frac{58.7 \times 10^{-4}}{15.09}$$

$$\therefore C_{o2} = 3.89 \times 10^{-4} \text{ in.}^5/\text{lb}$$

d. *Output inductance* (stage 2). The output inductance is defined as

$$L_o = \frac{\rho l}{A_{eff}}$$

The effective area of the output aperture and the length of the output aperture can be calculated from dimensions in Figure B.13. Directly,

$$l = 0.82 \text{ in.}$$

But

$$A_{eff} = \frac{A_{in} - A_{out}}{\ln \frac{A_{in}}{A_{out}}}$$

$$A_{in} = 4.7 \times 10^{-4} \text{ in.}^2$$

$$A_{out} = 18.8 \times 10^{-4} \text{ in.}^2$$

$$A_{eff} = \frac{4.7 \times 10^{-4} - 18.8 \times 10^{-4}}{\ln \frac{4.7 \times 10^{-4}}{18.8 \times 10^{-4}}}$$

$$A_{eff} = 1.01 \times 10^{-3} \text{ in.}^2$$

Then

$$\frac{l}{A_{eff}} = 8.09 \times 10^2 \text{ in.}^{-1}$$

208 Examples of V/Stol Control System Design

Similarly, for the output ferrule,

$$l = 0.44 \text{ in.}$$

$$A = 1.23 \times 10^{-2} \text{ in.}^2$$

$$\frac{l}{A} = 35.6 \text{ in.}^{-1}$$

Total $l/A = 809 + 35.6 = 845 \text{ in.}^{-1}$

The mass density of air at 1 atm is

$$\rho = 2.37 \times 10^{-3} \frac{\text{slugs}}{\text{ft}^3}$$

or

$$\rho = 1.15 \times 10^{-7} \frac{\text{lb sec}^2}{\text{in.}^4}$$

At the output bias pressure of stage 2 (15.09 psi absolute), the density is

$$\rho = 1.15 \times 10^{-7} \times \frac{15.09}{14.7}$$

$$\rho = 1.18 \times 10^{-7} \frac{\text{lb sec}^2}{\text{in.}^4}$$

Then

$$L_{o2} = \frac{\rho l}{A} = (1.18 \times 10^{-7})(8.45 \times 10^2)$$

$$L_{o2} = 9.93 \times 10^{-5} \frac{\text{lb sec}^2}{\text{in.}^5}$$

e. Input resistance (to stage 3). Input resistance is defined as

$$R_c = \frac{\Delta P_c}{\Delta Q_c}\bigg|_{(P_{c1} + P_{c2}) \text{ constant}}.$$

With reference to Figure B.19

$$R_{c3} = \frac{7 \times 10^{-2}}{1.08 \times 10^{-3}} \times \frac{60}{1728}$$

$$\therefore R_{c3} = 2.26 \frac{\text{lb sec}}{\text{in.}^5}.$$

f. Input capacitance (to stage 3). Input capacitance is defined as

$$C_c = \frac{V}{P_{abs}}$$

In the case of interconnected amplifiers, we have chosen to charge the line volume to the input of the amplifier. The amplifiers are connected together with 3 in. of $\frac{1}{8}$ ID tubing with a cross connection included for attaching shunt restrictors. Therefore, the total volume is made up of tubing, cross connector, ferrule connector, and amplifier input aperture. The volumes can be calculated from dimensions. The total input volume to stage 3 is

$$V = 0.0585 \text{ in.}^3$$

The pressure is the same as stage 2 output pressure; that is,

$$P = 0.39 \text{ psi gage} = 15.09 \text{ psi absolute}$$

Then

$$C_{c3} = \frac{0.0585}{15.09}$$

$$\therefore C_{c3} = 0.00389 \text{ in.}^5/\text{lb}$$

Note that

$$\text{at 10 rad/sec} \frac{1}{2\pi f C_{c3}} = \frac{1}{0.0389} = 25.7 \frac{\text{lb sec}}{\text{in.}^5} (1.6 \text{ Hz.})$$

$$\text{at 100 rad/sec} \frac{1}{2\pi f C_{c3}} = \frac{1}{0.389} = 2.57 \frac{\text{lb sec}}{\text{in.}^5} (16 \text{ Hz.})$$

g. Input inductance L_c. Input inductance is defined as

$$L_c = \frac{\rho l}{A_{eff}}$$

In the circuit shown in Figure B.30, we are charging all the inductance from the output ferrule of stage 2 into the control apertures of stage 3 as input inductance.

For the 3 in. of $\frac{1}{8}$ ID tubing,

$$\frac{l}{A} = 244 \text{ in.}^{-1}$$

For the cross connection,

$$\frac{l}{A} = 118 \text{ in.}^{-1}$$

For the control ferrule,

$$\frac{l}{A} = 35.6 \text{ in.}^{-1}$$

For the control aperture,

$$\frac{l}{A} = 131 \text{ in.}^{-1}$$

Total $l/a = 529$ in.$^{-1}$
Then

$$L_{c3} = 1.18 \times 10^{-7} \times 5.29 \times 10^2$$

$$\therefore L_{c3} = 6.21 \times 10^{-5} \frac{\text{lb sec}^2}{\text{in.}^5}$$

Note that

at 10 rad/sec $2\pi f L_{c3} = 6.21 \times 10^{-4} \frac{\text{lb sec}}{\text{in.}^5}$ (1.6 Hz)

at 100 rad/sec $2\pi f L_{ce} = 6.21 \times 10^{-3} \frac{\text{lb sec}}{\text{in.}^5}$ (16 Hz)

Comparing the inductive reactance of this circuit (which is typical of the yaw damper system) with the capacitive reactance under paragraph *f* shows that even up to a frequency of 100 rad/sec (16 Hz), the relative value of inductive reactance is small compared with the capacitive reactance. Since this circuit is typical of the yaw damper system, we have justified neglecting the inclusion of inductance in the derivation of the transfer functions.

h. Restrictors shunted to return. The value of the resistance shunted from the stage 2 output port to return can be calculated from Figure B.18. Assuming that it is a linear restrictor, its value can be calculated from the increase in flow that it provides at a given pressure. In Figure B.19 we have shown the addition of a shunt restrictor to return, which, at a pressure of 0.39 psi, raises the flow from 0.0062 to 0.0124 scfm. Then the resistance is

$$R_{g2} = \frac{0.39 \text{ psi}}{(0.0124 - 0.0062)} \times \frac{60}{1728}$$

$$\therefore R_{g2} = 2.16 \frac{\text{lb sec}}{\text{in.}^5}$$

i. Restrictors shunted line-to-line. The value of the resistance shunted across the output lines of the stage 2 amplifier can be calculated from

Figure B.19. Assuming a linear restrictor, its value can be calculated from the change in flow that it produces. In Figure B.19 the pressure swings 0.078 psi around the operating point. Without the line-to-line shunt, the flow swings 0.003 scfm. With the shunt, the flow swings 0.017 scfm. Therefore,

$$R_{s2} = \frac{0.078 \text{ psi}}{(0.017 - 0.003)} \times \frac{60}{1728}$$

$$\therefore R_{s2} = 0.194 \frac{\text{lb sec}}{\text{in.}^5}$$

j. Time delay. The time delay is made up of the transport delay in the second stage amplifier interaction chambers, plus wave propagation delays in the second stage output aperture and output ferrule, the tubing, the cross connectors, and the third stage input aperture and input ferrule.

In the stage 2 amplifier interaction chamber, the power jet exit velocity is

$$v_e = \frac{Q}{A} = \frac{0.026}{0.00025} \times \frac{1728}{60}$$

$$v_e = 2990 \text{ in./sec}$$

The receiver impact velocity (0.75 psi recovery) is

$$v_r = \sqrt{\frac{2p}{\rho}}$$

$$v_r = \sqrt{\frac{(6.44 \times 10^{-1})(0.745)(1.44 \times 10^2)}{8.47 \times 10^{-2}}}$$

$$v_r = 3420 \text{ in./sec}$$

Averaging gives

$$v \cong 3205 \text{ in./sec}$$

The length of the interaction chamber is 0.1 in. (see Figure B.13) Then,

$$t_1 = \frac{0.10}{3205}$$

$$t_1 = 3.1 \times 10^{-5} \text{ sec}$$

In the output aperture,

$$Q_0 = 1.26 \times 10^{-2} \text{ scfm}$$

Taking an average cross-sectional area 8.9×10^{-4},

$$v = \frac{1.26 \times 10^{-2}}{8.9 \times 10^{-4}} \times \frac{1728}{60}$$

$$v = 407 \text{ in./sec}$$

The length of the output aperture is approximately 0.82 in. Then,

$$t_2 = \frac{0.82}{407 + 13200}$$

where 13200 in./sec is the propagation velocity of sound

$$t_2 = 6.03 \times 10^{-5} \text{ sec}$$

In the output ferrule,

$$v = \frac{1.26 \times 10^{-2}}{1.23 \times 10^{-2}} \times \frac{1728}{60}$$

$$v = 29.5 \text{ in./sec}$$

The length of the ferrule is 0.44 in. Then,

$$t_3 = \frac{0.44}{29.5 + 13200}$$

$$t_3 = 3.32 \times 10^{-5} \text{ sec}$$

Similarly, we calculate the time delay in the tubing,

$$t_4 = 2.28 \times 10^{-4} \text{ sec}$$

and in the cross connector,

$$t_5 = 6.14 \times 10^{-5} \text{ sec}$$

and in the control ferrule,

$$t_6 = 3.32 \times 10^{-5} \text{ sec}$$

and in the control aperture,

$$t_7 = 2.80 \times 10^{-5} \text{ sec}$$

Adding up the individual time delays, t_1 through t_7, we have

$$t_{23} = 4.75 \times 10^{-4} \text{ sec}$$

k. Circuit parameters. A listing of all equivalent circuit parameters for the fluidic yaw damper is given in Table B.1.

17. Substitute the Performance Parameters in the Transfer Functions. With reference to the overall system transfer curve shown in Figure B.24 we have shown that the steady-state gain requirements of the fluidic yaw damper have been met. Therefore, in calculating the dynamic behavior, we will consider only the frequency-sensitive portions of the transfer functions.

Referring now to Figure B.29,

a.
$$\left[\frac{P_{cd1}}{\omega}\right] = \left[1 + s(C_{o\omega} + C_{c1}) \frac{R_{c1}R_{o\omega}R_{s\omega}}{R_{c1}R_{o\omega} + R_{o\omega}R_{s\omega} + R_{c1}R_{s\omega}}\right]^{-1}$$

$$\left[\frac{P_{cd1}}{\omega}\right] = \left[1 + s(0.0008 + 0.0033) \frac{0.348 \times 0.435 \times 0.057}{0.348 \times 0.435 + 0.435 \times 0.057 + 0.348 \times 0.057}\right]^{-1}$$

$$\left[\frac{P_{cd1}}{\omega}\right] = \frac{1}{1 + 0.00018s}$$

b.
$$\left[\frac{P_{cd2}}{P_{cd1}}\right] = \left[1 + s(C_{o1} + C_{c2}) \frac{R_{c2}R_{o2}R_{sa}}{R_{c2}R_{o1} + R_{o1}R_{sa} + R_{c2}R_{sa}}\right]^{-1}$$

Table B.1 Tabulation of Equivalent Circuit Parameters[a]

$R_{o\omega} = 0.435$	$R_{o3} = 2.22$
$R_{c1} = 0.348$	$R_{c4} = 4.00$
$R_{s\omega} = 0.057$	$R_{s3} = 0.343$
$C_{o\omega} = 0.0008$	$R_{g3} = 4.49$
$C_{c1} = 0.0033$	$C_{o3} = 0.00038$
$R_{o1} = 0.668$	$C_{c4} = 0.00526/0.0005$
$R_{c2} = 0.652$	$R_{o4} = 3.22$
$R_{s1} = 0.142$	$R_{c5} = 5.00$
$R_{g1} = 2.11$	$R_{s4} = 0.393$
$C_{o1} = 0.0004$	$R_{g4} = 1.50$
$C_{c2} = 0.00394$	$C_{o4} = 0.00038$
$R_{o2} = 1.29$	$C_{c5} = 0.0053/0.0005$
$C_{o2} = 0.00038$	$R_{o5} = 3.34$
$R_{c3} = 2.26$	$R_{g5} = 8.43$
$R_{g2} = 2.16$	$C_{o5} = 0.00032$
$R_{s2} = 0.194$	$C_{L} = 0.0020/0.0010$
$C_{c3} = 0.00389$	total $t = 0.054/0.035$

[a]Note: units of R, lb sec^2/in^5; units of C, in^5/lb.

where
$$R_{sa} = \frac{R_{s1}R_{g1}}{R_{s1}+R_{g1}}$$

$$\left[\frac{P_{cd2}}{P_{cd1}}\right] = (1+0.0004s)^{-1}$$

c.
$$\left[\frac{P_{cd3}}{P_{cd2}}\right] = \left[1+s(C_{o2}+C_{c3})\frac{R_{c3}R_{o2}R_{sb}}{R_{c3}R_{o2}+R_{o2}R_{sb}+R_{c3}R_{sb}}\right]^{-1}$$

where
$$R_{sb} = \frac{R_{s2}R_{g2}}{R_{s2}+R_{g2}}$$

$$\left[\frac{P_{cd3}}{P_{cd2}}\right] = (1+0.0006s)^{-1}$$

d.
$$Z_4 = R_{c4}[1.5+(1+sC_{c4}R_{c4})^{-1}]$$

$$Z_4 = \frac{10+0.126s}{1+0.021s}$$

e.
$$Z_5 = R_{c4}[1.5+(1+sC_xR_{c4})^{-1}]$$

Assuming that we want $P_{cd4b}/P_x = 1/1+3s$ to provide the proper highpass network time constant,
$$Z_5 = \frac{10+30s}{1+5s}$$

f.
$$Z_p = \frac{Z_4Z_5}{Z_4+Z_5}$$

$$Z_p = \frac{100+301s+3.78s^2}{20+80.3s+1.26s^2}$$

g.
$$\left[\frac{P_x}{P_{cd3}}\right] = \left(1+sC_{o3}\frac{R_{o3}Z_pR_{g3}}{R_{o3}R_{g3}+R_{o3}Z_p+R_{g3}Z_p}\right)^{-1}$$

$$\left[\frac{P_x}{P_{cd3}}\right] = (1+0.00043s)^{-1} \text{ (max)}$$

h.
$$\left[\frac{P_{cd4a}}{P_x}\right] = (1+0.6sC_{c4}R_{c4})^{-1}$$

$$\left[\frac{P_{cd4a}}{P_x}\right] = (1+0.0126s)^{-1}$$

i.
$$Z_a = 2R_{c5} + \frac{R_{o4}}{1+sC_{o4}R_{o4}}$$

$$Z_a = \frac{13.2+0.012s}{1+0.0012s}$$

j.
$$Z_b = \frac{Z_a R_{c5}}{sC_{c5}Z_a R_{c5} + R_{c5} + Z_a}$$

$$Z_b = \frac{3.52}{1+0.019s}$$

k.
$$Z_c = 2R_{c5} + Z_b$$

$$Z_c = \frac{13.5+0.19s}{1+0.019s}$$

l.
$$\left[\frac{P_a}{P_{cd4a}}\right] = \left(1 + sC_{o4}\frac{Z_c R_{g4} R_{o4}}{R_{o4}R_{g4} + Z_c R_{o4} + Z_c R_{g4}}\right)^{-1}$$

$$\left[\frac{P_a}{P_{cd4a}}\right] = (1+0.00036s)^{-1} \text{ (max)}$$

m.
$$\left[\frac{P_{cd5}}{P_a}\right] = \left(1 + sC_{c5}\frac{2R_{c5}Z_a}{2R_{c5}+3Z_a}\right)^{-1}$$

$$\left[\frac{P_{cd5}}{P_a}\right] = (1+0.014s)^{-1} \text{ (max)}$$

n.
$$\left[\frac{P_{od}}{P_{cd5}}\right] = \left[1 + s(C_{o5}+C_L)\frac{R_{o5}R_{g5}}{R_{g5}+R_{o5}}\right]^{-1}$$

$$\left[\frac{P_{od}}{P_{cd5}}\right] = (1+0.0055s)^{-1}$$

o.
$$\left[\frac{P_{cd4b}}{P_x}\right] = (1+0.6sC_x R_{c4})^{-1}$$

$$\left[\frac{P_{cd4b}}{P_x}\right] = (1+3.0s)^{-1}$$

p.
$$\left[\frac{P_b}{P_{cd4b}}\right] = \left(1 + sC_{o4}\frac{Z_c R_{g4} R_{o4}}{R_{o4}R_{g4}+Z_c R_{o4}+Z_c R_{g4}}\right)^{-1}$$

$$\left[\frac{P_b}{P_{cd4b}}\right] = (1+0.00036s)^{-1} \text{ (max)}$$

216 Examples of V/Stol Control System Design

q.
$$\left[\frac{P_{cd5}}{P_b}\right] = \left(1 + sC_{c5}\frac{2R_{c5}Z_a}{2R_{c5}+3Z_a}\right)^{-1}$$

$$\left[\frac{P_{cd5}}{P_b}\right] = (1+0.014s)^{-1} \text{ (max)}$$

With reference to Figure B.28 the overall frequency-sensitive portion of the transfer function is

$$\frac{P_{od}}{\omega} = \frac{P_{cd1}}{\omega} \cdot \frac{P_{cd2}}{P_{cd1}} \cdot \frac{P_{cd3}}{P_{cd2}} \cdot \frac{P_x}{P_{cd3}} \left[\frac{P_{cd4a}}{P_x} \cdot \frac{P_a}{P_{cd4a}} \cdot \frac{P_{cd5}}{P_a} - \frac{P_{cd4b}}{P_x} \cdot \frac{P_b}{P_{cd4b}} \cdot \frac{P_{cd5}}{P_b}\right] \frac{P_{od}}{P_{cd5}}$$

The frequency response of the individual transfer functions will be evaluated separately and the results analyzed; then they will be combined according to the formula above.

18. Calculate the System Frequency Response. As described in Chapter 9, the frequency response is calculated by substituting $j2\pi f$ for s in the transfer function. The phase shift due to time delay is evaluated as a separate factor, and is then added to the total transfer function.

Substituting $j2\pi f$ for s at $f = 10$ Hz,

$(2\pi f = 62.8$ rad/sec$)$

$$\left[\frac{P_{cd1}}{\omega}\right] = (1+0.00018s)^{-1} = (1+j0.00018 \times 62.8)^{-1} = (1+j0.0113)^{-1}$$

$$\left[\frac{P_{cd1}}{\omega}\right] = 1\underline{/-0.5°}$$

$$\left[\frac{P_{cd2}}{P_{cd1}}\right] = (1+0.0004s)^{-1} = (1+j0.0004 \times 62.8)^{-1} = (1+j0.025)^{-1}$$

$$\left[\frac{P_{cd2}}{P_{cd1}}\right] = 1\underline{/-1.5°}$$

$$\left[\frac{P_{cd3}}{P_{cd2}}\right] = (1+0.0006s)^{-1} = (1+j0.0006 \times 62.8)^{-1} = (1+j0.0375)^{-1}$$

$$\left[\frac{P_{cd3}}{P_{cd2}}\right] = 1\underline{/-2°}$$

$$\left[\frac{P_x}{P_{cd3}}\right] = (1+0.00043s)^{-1} = (1+j0.00043 \times 62.8)^{-1} = (1+j0.027)^{-1}$$

$$\left[\frac{P_x}{P_{cd3}}\right] = 1\underline{/-1.5°}$$

$$\left[\frac{P_{cd4a}}{P_x}\right] = (1+0.0126s)^{-1} = (1+j0.0126 \times 62.8)^{-1} = (1+j0.79)^{-1}$$

$$\left[\frac{P_{cd4a}}{P_x}\right] = 0.78 \underline{/-38°}$$

$$\left[\frac{P_a}{P_{cd4a}}\right] = (1+0.00036s)^{-1} = (1+j0.00036 \times 62.8)^{-1} = (1+j0.0225)^{-1}$$

$$\left[\frac{P_a}{P_{cd4a}}\right] = 1 \underline{/-1.5°}$$

$$\left[\frac{P_{cd5}}{P_a}\right] = (1+0.014s)^{-1} = (1+j0.014 \times 62.8)^{-1} = (1+j0.88)^{-1}$$

$$\left[\frac{P_{cd5}}{P_a}\right] = 0.76 \underline{/-41°}$$

$$\left[\frac{P_{od}}{P_{cd5}}\right] = (1+0.0055s)^{-1} = (1+j0.0055 \times 62.8)^{-1} = (1+j0.35)^{-1}$$

$$\left[\frac{P_{od}}{P_{cd5}}\right] = 0.95 \underline{/-19°}$$

$$\left[\frac{P_{cd4b}}{P_x}\right] = (1+3s)^{-1} = (1+j3 \times 62.8)^{-1} = (1+j190)^{-1}$$

$$\left[\frac{P_{cd4b}}{P_x}\right] = 0.0053 \underline{/-90°}$$

$$\left[\frac{P_b}{P_{cd4b}}\right] = 1 \underline{/-1.5°}$$

$$\left[\frac{P_{cd5}}{P_b}\right] = 0.76 \underline{/-41°}$$

The phase shift due to time delay is

$$\theta_d = 360\, ft_d = 360 \times 10 \times 0.054$$

$$\theta_d = 194°$$

19. Plot the Frequency Response of the System and Compare It with Specified requirements. The values of the transfer functions at 10 cps are shown in their appropriate places in Figure B.31. The lower loop through the high pass network is highly attenuated, so it will be neglected in the evaluation of the yaw damper at high frequencies. Adding up the total phase

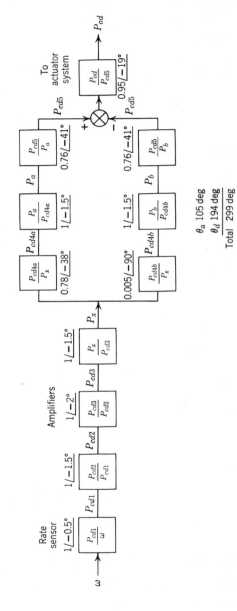

Figure B.31 Preliminary values of the transfer functions at 10 Hz.

shift in the system transfer functions through the upper loop, we have

$$\theta_a = 105°$$

Adding

$$\theta_d = 194°$$

gives a total

$$\theta = 299°$$

for the complete system. Comparing this with the specified allowable phase shift (180°) shows that *the requirements at 10 Hz have not been met.*

20. Investigate Individual Component Transfer Functions and Make Necessary Changes. Examination of the phase shifts shown in Figure B.31 reveals that the major contributions come from:

1. The input circuit of the high pass network (stage 4).
2. The input circuit to the summing amplifier (stage 5).
3. The output circuit of the amplifier driving the actuator (stage 5).

Further review of the transfer functions related to stages 4 and 5 shows that the input capacitance to stage 4 (C_{c4}) the input capacitance to stage 5 (C_{c5}), and the load circuit capacitance on stage 5 (C_L) can be reduced as follows.

The input circuit to stage 4 contains series resistance; therefore, we can reduce the ID of the tubing to a very small value, thereby providing some of the required resistance while reducing the trapped volume of air a great degree. In re-examining the physical circuit layout, we find that it is also possible to reduce the length of tubing. Therefore, it is practical to reduce the value of C_{c4} by a factor of 10.

The input circuit to stage 5 also contains series resistance and excessive lengths of tubing. By the same reasoning applied above, we find it practical to reduce the value of C_{c5} by a factor of 10.

In the load circuit of stage 5, the total capacitance (0.002 in.^{5}/lb) is primarily due to tubing (the actuator is approximately 0.0005 in.^{5}/lb). The actuator has infinite input resistance, so the tubing running to it can be very small without introducing a significant amount of loss. Therefore, it appears practical that the total load circuit capacitance can be reduced to less than 0.0008 in.^{5}/lb.

Using these new lengths and diameters of tubing also reduces the time delay, because of shorter distances of signal travel and higher flow velocities. Recalculating the total time delay based on new networks of tubing results in a total delay of 0.035 sec.

18. Calculate the System Frequency Response. Recalculating the transfer functions affected by these changes, we have

$$\left[\frac{P_{cd4a}}{P_x}\right] = (1 + 0.6sC'_{c4}R_{c4})^{-1} = (1 + 0.6s \times 0.0005 \times 4)^{-1}$$

$$\left[\frac{P_{cd4a}}{P_x}\right] = (1 + 0.00126s)^{-1}$$

at 10 Hz

$$\left[\frac{P_{cd4a}}{P_x}\right] = (1 + j0.00126 \times 62.8)^{-1} = (1 + j0.079)^{-1}$$

$$\left[\frac{P_{cd4a}}{P_x}\right] = 1\underline{/-4.5°}$$

$$\left[\frac{P_{cd5}}{P_a}\right] = \left(1 + sC'_{c5}\frac{2R_{c5}Z_a}{2R_{c5} + 3Z_a}\right)^{-1}$$

$$\left[\frac{P_{cd5}}{P_a}\right] = (1 + 0.0014s)^{-1}$$

at 10 Hz

$$\left[\frac{P_{cd5}}{P_a}\right] = (1 + j0.0014 \times 62.8)^{-1}(1 + j0.088)^{-1}$$

$$\left[\frac{P_{cd5}}{P_a}\right] = 1\underline{/-5°}$$

$$\left[\frac{P_{od}}{P_{cd5}}\right] = \left[1 + s(C_{o5} + C'_c)\frac{R_{o5}R_{g5}}{R_{g5} + R_{o5}}\right]^{-1}$$

$$\left[\frac{P_{od}}{P_{cd5}}\right] = (1 + 0.0026s)^{-1}$$

at 10 Hz

$$\left[\frac{P_{od}}{P_{cd5}}\right] = (1 + j0.0026 \times 62.8) = (1 + j0.16)^{-1}$$

$$\left[\frac{P_{od}}{P_{cd5}}\right] = 0.98\underline{/-9°}$$

The phase shift due to time delay at 10 Hz is now

$$\theta_d = 360\, ft'_d = 360 \times 10 \times 0.035$$

$$\theta_d = 126°$$

Referring to Figure B.32, which contains the final values for the transfer functions of the individual blocks at 10 Hz, we see now that the total phase shift is 152°, *well within the specified maximum of 180°*.

Since we have shown that there is a little phase shift in the individual circuits (except P_{cd4b}/P_x) of the fluidic yaw damper at frequencies below 10 Hz, the only circuit we are concerned with at other frequencies is the parallel loop in the high pass network and the difference

$$\left[\frac{P_{od}}{\omega}\right] \cong \left[\frac{P_{cd4a}}{P_x}\right] - \left[\frac{P_{cd4b}}{P_x}\right]$$

The calculated values are tabulated below.

f_{Hz}	$\left[\dfrac{P_{cd4a}}{P_x}\right]$	$\left[\dfrac{P_{cd4b}}{P_x}\right]$	$\left[\dfrac{P_{od}}{\omega}\right]$	$\left[\dfrac{P_{od}}{\omega}\right] db$
0.001	1 /0	1 /−1	0.017 /+90	−34db /+90
0.01	1 /0	0.98 /−11	0.195 /+78	−14db /+78
0.1	1 /0	0.47 /−62	0.85 /+28	−1.4db /+28
1.0	1 /0	0.053 /−87	0.997 /+3	0db /+3
10	1 /−4.5	0.0053 /−90	1 /−4	0db /−4

To include the effect of time delay, the following phase shifts must be added:

f_{Hz}	θ_d
10.0	126.0
1.0	12.6
0.1	1.26
0.01	0.13
0.001	0.013

19. Plot the Frequency Response of the System and Compare It with Specified Requirements. The results are plotted as a Bode diagram in Figure B.33 and compared with the specified frequency response. The fluidic yaw damper has been designed to meet the necessary performance requirements with a reasonable margin of safety.

21. Finalize Preliminary Design

22. List Factors Important to the Performance of the System. The volume required to generate the high pass network time constant of 3.0 seconds can be calculated from the expression P_{cd4a}/P_x (see transfer function

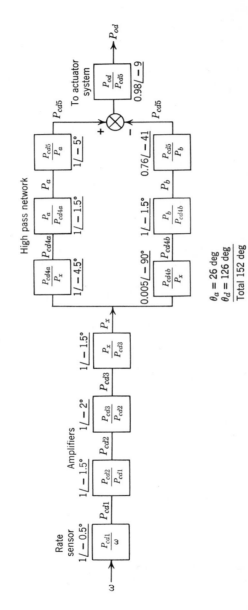

Figure B.32 Final values of the transfer functions at 10 Hz.

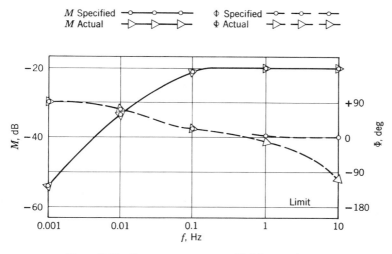

Figure B.33 Frequency response of fluidic yaw damper.

listed in Figure B.29). In that equation the effective time constant is $0.6C_x R_{c4}$. Then,

$$0.6 C_x R_{c4} = 3.0 \text{ sec}$$

$$C_x = \frac{3.0}{0.6 R_{c4}}$$

Substituting the value of $R_{c4} = 4.0$.

$$C_x = \frac{3.0}{0.6 \times 4.0}$$
$$C_x = 1.25 \text{ in.}^5/\text{lb}$$

Then, since

$$C_x = \frac{V_x}{P_{\text{abs}}}$$

and P_{abs} in the circuit is 15.13 psia (0.43 psi gage),

$$V_x = C_x P_{\text{abs}} = 1.25 \times 15.13$$
$$V_x = 19.2 \text{ in.}^3$$

That is, to obtain the proper high pass network time constant, we must add volume of 19.2 in.³ to each fourth stage input network in the position (C_x) shown in Figure B.26.

For most linear operation at all frequencies of interest, it is important to design the two parallel paths in the high pass network with identical

224 Examples of V/Stol Control System Design

characteristics. Otherwise, there would be more or less phase shift in one path than the other, and the summed output from stage 5 would be seriously distorted.

23. Design Summary. In summary, we have designed a fluidic yaw damper to meet certain performance requirements using available fluidic components. A schematic of the system is shown in Figure B.23 and the static transfer characteristics are illustrated in Figure B.24. The equivalent electrical circuit is shown in Figure B.25 a transfer function block diagram is shown in Figure B.28 and a listing of the corresponding transfer functions is shown in Figure B.29. The parameters of the transfer functions are listed in Table B.1 and the final frequency response shown in Figure B.33 is compared with the specified requirements.

References

1. *Aerospace Recommended Practice-ARP 993-Fluidic Technology*, Society of Automotive Engineers, Inc., New York, N.Y.
2. Belsterling, C. A., "Fluidic Systems Design Manual," *USAAVLABS Report 67-32*, U.S. Army Aviation Material Laboratories, Fort Eustis, Va., May 1967.
3. Belsterling, C. A., "Development of the Techniques for the Static and Dynamic Analysis of Fluid State Components and Systems," *USAAVLABS Report 66-16*, U.S. Aviation Materiel Laboratories, Fort Eustis, Va., February, 1966.
4. Belsterling, C. A., "A Systematic Approach for Designing Fluidic Systems," *Control Eng.*, April 1966.
5. Belsterling, C. A., and K. C. Tsui, "Analyzing Proportional Fluid Amplifier Circuits," *Control Eng.*, August 1965.
6. Belsterling, C. A., and K. G. Tsui, "Application Techniques for Proportional Pure fluid Amplifiers," *Proc. Second Fluid Amplification Symp.*, Harry Diamond Laboratories, Washington, D. C., May 1964.
7. Blackburn, J. F., G. Reethof, and J. L. Shearer, "Fluid Power Control," The Technology Press of *M. I. T.* and John Wiley & Sons, Inc., New York and London, 1960.
8. Boothe, W. A., "A Lumped Parameter Technique for Predicting Analog Fluid Amplifier Dynamics," *Proc. Joint Automatic Control Conference*, Stanford, Calif., June 1964.
9. Chestnut, H., and R. W. Mayer, "Servomechanisms and Regulating System Design," Vol. 1, John Wiley & Sons, Inc., New York, 1951.
10. Fox, H. L., and O. Lew Wood, "The Development of Basic Devices and the Need for Theory," *Control Eng.*, Special Report, October 1964.
11. Goldman, S., *Transformation Calculus and Electrical Transients*, Prentice-Hall, Inc., New York, 1949.
12. Guillemin, E. A., *The Mathematics of Circuit Analysis*, John Wiley & Sons, Inc., New York, 1949.
13. Katz, S., and R. J. Dockery, "Staging of Proportional and Bistable Fluid Amplifiers," *Report TR-1165*, Harry Diamond Laboratories, Washington, D.C., August 1963.

References

14. Kirshner, J. M., Ed., *Fluid Amplifiers*, McGraw-Hill Book Company, Inc., New York, 1966.
15. Letham, D. L., "Stream Interaction Amplifiers," *Mach. Des.*, June 23, 1966.
16. Letham, D. L., "Turbulence Amplifiers," *Mach. Des.*, July 7, 1966.
17. Letham, D. L., "Vortex Amplifiers," *Mach. Des.*, July 21, 1966.
18. Letham, D. L., Impact Modulators," *Mach. Des.*, August 4, 1966.
19. Letham, D. L., "Miscellaneous Devices," *Mach. Des.*, August 18, 1966.
20. Mayer, E. A., and P. Maker, "Control Characteristics of Vortex Valves," *Proc. Second Fluid Amplification Symposium*, Harry Diamond Laboratories, Washington, D.C., May 1964.
21. Ryder, J. D., *Electronic Engineering Principles*, Prentice-Hall, Inc., New York, 1947.
22. Shinn, J. N., and W. A. Boothe, "Connecting Elements Into Circuits and Systems," *Control Eng.*, Special Report, October 1964.
23. Shinners, S. M., "Fluidics," *Electrotechnol.*, March 1967.
24. Thorne, G. H., "Fluidic Essentials," publication of the Fluidonics Division, Imperial Eastman Corp.
25. Wood, O. Lew, "Pure Fluid Devices," *Mach. Des.*, June 24, 1965.

Index

Acoustic beam sensor, 68
Active, definition of, 157
Actuator, definition of, 162
  descriptions of, 26
  equivalent circuit, 141
Admittance, parameter, 16
  reverse transfer, 26
  vacuum triode, 11
Admittance matrix, 18, 20, 39
  measurement of, 26
Air motors, equivalent circuit, 141, 142
Amplification factor, calculation of, 149
  pressure, 17
  vacuum triode, 4
Amplifier, axisymmetric focussed jet, 62–64
  boundary layer control, 50–52
  definition of, 158
  description of, 25
  impact modulator, 52, 53
  jet interaction, 44–48
  turbulence, 61–63
  vortex, 48–50
  wall attachment, 59–61
Analog, computer simulation, 23
  definition of, 157
  systems design, 25
Analogies, electrical, mechanical and hydraulic, 14, 15
  fluid to electric, 16, 21
Analogous, circuits, 3
  electric circuits, 19
  elements, acoustical, 7
Analogy, flow to current, 15
  pressure to voltage, 15
Analysis, circuits, 22
  coupled networks, 13
  dynamic, 22, 69
  static, 22, 69
Analytical method, optimum, 3
ASME Fluidics Symposium, 25
Axisymmetric focused jet amplifier, definition of, 160
  description of, 62–64

Back pressure sensor, 53, 54
Beam deflection amplifier, 23
Belsterling, C. A., 18, 22, 23, 25
Bias point, definition of, 158
Bibliography, chronological, 27–30
Bistable amplifier, 18
  analysis of, 19
Black box, 3, 17, 18
Bleed port, matching, 16
Bode diagram, 95, 96
Bondgraphs, 21
Booster amplifiers, 33
Boothe, W. A., 17, 19, 20
Boundary layer control amplifier, definition of, 158
  description of, 50–52
Bowles, R. E., 15
Brown, F. T., 3, 17, 18, 21, 26
Bubbler tube sensor, 55

Capacitance, adiabatic, 15
  equivalent, 82
  hydraulic, 10
  isothermal, 15
  pneumatic, 8
Capacitor, definition of, 163
Cascaded amplifiers, 18
Cascaded fluidic devices, 20
Cascaded sensor and amplifier, 143–153
Cascaded stages, bistable, 18
Cascading fluidic amplifiers, 16

227

Index

Cascading impact modulators, 19
Characteristic curves, 18
  normalization of, 75
Characteristics, definition of, input, 69
    output, 69
    transfer, 69
  bistable amplifiers, 23
    dynamic, 36
    input, 34–36
    output, 35–37
    proportional amplifiers, 23
    supply, 35, 37
    transfer, 17
    vacuum triode, 4
Circuit analysis, nonlinear, 13
Circuit synthesis, 17
Closed jet interaction amplifier, 22
Component, definition of, 157
Component description, 13, 14, 17, 25
  analogous circuit, 3
  coupled equations, 3
  graphical, 3
  model, 3
Component layout, 33
Component requirements, 31
Component selection, 31, 39
Compressibility, 20
Converging jet sensor, 56
Coupled equations, 3
Coupled multiport elements, 18
Coupled networks, 13
Current analog, 22

Design, analog systems, 25
  checklist, 156
  process, 31
  summary, 224
  techniques, graphical, 19
Dexter, E. M., 16, 17
Diagram, block, 33
  functional, 33, 34
Diamond Ordnance Fuze Laboratories, 15
Digital, definition of, 157
Displays, definition of, 162
Distributed parameter networks, hydraulic 10
Diverging jet sensor, 56, 57
  long range, 67, 68
Dockery, R. J., 17, 19
Dynamic analysis, 18, 20, 23, 40, 41, 70
  vacuum triode, 6
Dynamic characteristics, 21
Dynamic performance, limits on, 21
Dynamic response, prediction of, 19
Dynamics, measurement of, 26
  receiver and load, 20

Effective area, 78
Effective capacitance, pneumatic, 8
Effective volume, 78
Elbow amplifier, equivalent circuit, 138, 139
Electon tube technology, 4
Electrical equivalent circuits, 22
Electric analogies, 21
Elements, definition of, 157
Empirical parameters, 12
Environment, 32
  explosive, 33
  high temperature, 32
  moist and oily, 33
Equivalent capacitance, definition of, 82
Equivalent circuit, acoustical, 7
  analog amplifier, 76
  closed jet interaction amplifier, 22, 137, 138
  coupled system, 199, 200
  development of, 133–136
  digital amplifier, 77
  human vocal tract, 7, 9
  hydraulic, 10
  hydraulic-mechanical, 10, 11
  piston-type actuator, 141
  piston-type and vane-type motors, 141, 142
  pneumatic, 15
  proportional amplifier, 22, 37, 38
  transistor, 12
  vacuum triode, 6, 7
  vented elbow amplifier, 138, 139
  vented jet interaction amplifier, 136, 137
  vented wall attachment amplifier, 139, 140
  vortex rate sensor, 140, 141
Equivalent circuit analysis, 18
  mechanics, 8
  pneumatics, 8
  vacuum triode, 4, 6
Equivalent circuit design, logic circuits, 19
Equivalent circuit methods, 21
Equivalent circuit parameters, calculation of, 205, 213
  transistor, 12
  vacuum triode, 6

Equivalent electrical circuits, 21, 23, 25
Equivalent inductance, definition of, 82
Experimental techniques, 26
Ezekiel, F. D., 15

Failure characteristics, 43
First Fluid Amplification Symposium, 16
Flip-flop, definition of, 162
　wall attachment, 35
Flow gain, definition of, 81
Flow transducers, 101, 102
Flueric, definition of, 157
Fluid Amplification Symposium, First, 16
　; Second, 18
　Third, 22
Fluid Amplifiers (book), 23
Fluid circuit theory, 23
Fluid-electric analogy, 22
Fluid flow equations, 20
Fluidics, book, 20
　definition of, 157
Fluidics technology, review of, 14
Fluidic yaw damper, 25
Fluid transmission lines, 23
Four terminal network, 17
Fox, H. L., 20, 21
Frequency response, calculation of, 216, 217, 220, 221
　effect of supply pressure, 42, 43
　fluidic yaw damper, 223
　jet interaction amplifier, 22
Frequency response testing, 95
Frequency response tests, air, 19
　water, 19
Frequency spectrum, acoustic frequencies, 2
　band 1, 1
　band 2, 1
　band 3, 2
　fluidic systems, 1
　higher frequencies, 1
　static and low, 1
Functional models, 12
Future of analysis, 21

Gain, 32
　measurement of, 95
　pressure, 17
General Electric Company, 20
Goldschmeid, R. F., 21
Goto, J. L., 19

Graphical analysis, 3, 40
　fluid transmission, 10
　transistor, 12
Graphical characteristics, 25
　actuators, 114
　back pressure sensor, 112
　closed jet interaction amplifier, 106, 108
　fluid motors, 114
　interruptable jet sensor, 112
　laminar restrictors, 104, 105
　turbulence amplifier, 110, 111
　turbulent restrictors, 105
　vented elbow amplifier, 108
　vented jet interaction amplifier, 106, 107
　vented wall attachment flip-flop, 109, 110
　vortex rate sensor, 113
Graphical design, 19, 22, 23
　bistable amplifiers, 19
　load line methods, 20
　logic circuits, 19
Graphical matching, 17
Graphical output characteristics, 17, 19
Grogan, E. C., 15

Harry Diamond Laboratories, 23
Hicks, B. A., 16
Horton, B., 15
Humphrey, R. A., 26
Hydraulic-mechanical models, 10
Hydraulics technology, 9

Iberall, A. S., 14
Impact modulator, 52, 53
　cascading, 19
　definition of, 160
Impedance, definition of, 158, 162
　dynamic measurement of, 99, 100
　input, 16, 17, 19
　　vacuum triode, 11
　matching, 16
　output, 16
Inductance, equivalent, 82
Inductor, definition of, 163
Inertance, 20
　hydraulic, 10
Input capacitance, calculation of, 149
Input characteristics, 17
　analog amplifiers, 70, 71
　definition of, 69
　digital amplifiers, 73, 74

Input impedance, 16, 17, 19
　vacuum triode, 11
Input inductance, calculation of, 149
Input resistance, calculation of, 149
　definition of, 81
Integrated circuits, 33
Interface, 32
Internal feedback, 22
Interruptable jet, laminar, 66, 67
　turbulent, 54, 55

Jet interaction amplifier, closed, 22, 46–48
　definition of, 158
　equivalent circuits, 136–138
　vented, 18, 34, 44–46
Jet path length, 78
Jetter, E. S., 16
Joint Automatic Control Conference, 19

Karam, J. T., 23
Katz, S., 16, 17, 19, 22
Kirshner, J. M., 22, 23

Laplace transform, 150
Large signal analysis, vacuum triode, 4
Lechner, T. J., 16, 19
Limit valve sensors, 65, 66
Linear circuit, transistors, 12, 13
Linear equivalent circuits, 20
Line lengths, 77
Line resistance, calculation of, 149
Load line, 19
　performance analysis, 115, 119
　technique, 18, 23, 40
　vacuum triode, 4
Loading, definition of, 158
Load resistance, calculation of, 150
Logic, 32
Logic device, 64, 65
　definition of, 162
Long interconnections, 33
Lumped parameters, 21
　acoustic models, 7
　representations, 19
　system, 20

Maker, E. A., 19
Manion, F. M., 25
Matching, 23
　cascaded stages, 16
　digital amplifier and flip-flops, 127–132
　graphically, 17
　procedure, 18, 23
　sensor and amplifier, 122–127, 132
Matching bias points, 183–186, 187–190, 194–197
　analog, 125, 126
　digital, 129, 130
Matching operating ranges, analog, 186–190, 192–197
　digital, 130, 131
Mathematical approaches, 21
　description, 41, 42
　representations, 21
Matrices, admittance, 18
　self admittance, 26
　transfer admittance, 26
Matrix, admittance, 39
　analysis, 41, 42
　describers, 38, 39
　elements, measurement of, 26
Mayer, P., 19
Measuring, admittance matrices, 26
　gain, 95
　matrix elements, 26
　phase angle, 94, 95
　switching time, 97
　time delay, 95, 96
Measuring dynamics, 26
　of passive elements, 99, 100
　of sensors, 98, 99
Models, 3
　dynamic, 21
　functional, 12
　hydraulic-mechanical, 10
　lumped parameter, 7
　lumped pure delay, 21
　multiport, 16
　physical, 12
　transistor, 11, 12
Motors, equivalent circuit, 141, 142
Multiport models, 11, 16
　networks, 18, 26

NASA, 20
Networks, two-port, 15
Nomenclature, 163, 164
Nonlinear circuit analysis, 13
Nonlinear systems analysis, analog computers, 13

Norwood, R. D., 16
n-port network, 20

Operating fluid, qualities of, 79
Operating range, definition of, 122, 158
Oscillatory pressures, transmission of, 14
Oscilloscope, 102
  displays, 94, 95
  dynamic testing, 93, 94
Output capacitance, calculation of, 148, 150
Output characteristics, 17–19
  analog amplifier, 71, 72
  definition of, 69
  digital amplifier, 74, 75
  graphical, 17
  vortex valve, 17
Output impedance, 16
Output inductance, calculation of, 148, 150
Output resistance, calculation of, 147–149
  definition of, 79

Parameters, empirical, 12
  transistor, 11, 12
Parameter studies, 43
  definition of, 32
Passive, definition of, 158
Passive AND gate, 64, 65
Passive logic devices, 64, 65
Paynter, H. M., 15
Performance analysis, definition of, 32
Performance parameters, 19
Physical models, 12
Physical systems, 39
Pill, Juri, 26
Piecewise linear analysis, 13
Plate characteristics, vacuum triode, 5
Plate resistance, vacuum triode, 4
Pneumatic signal generator, 101
Pneumatic technology, 7
Pneumatic transmission lines, 14
Power jet velocity, 78
Power nozzle characteristics, 75
Pressure amplification factor, 17
  definition of, 80
Pressure gain, 17
  definition of, 79, 80
Pressure recovery factor, definition of, 83
Pressure transducers, 93, 101, 102
Proportional amplifier, 18
Pure-delay models, 21

Quiescent point, definition of, 158

Random testing technique, 26
Rate sensor, equivalent circuit, 140, 141
  vortex, 58, 59
References, 225, 226
Resistance, hydraulic, 10
Resistor, definition of, 162
Response, 26
Restrictor characteristics, 104, 105
Reverse transfer admittance, 20, 22, 26
  vacuum triode, 11
Roffman, G. L., 22
Rohman, C. P., 15

Saghati, H. T., 19
Scattering variables, 18
Second Fluid Amplification Symposium, 18
Self admittance matrices, 26
Sensitivity factor, calculation of, 147, 148
  definition of, 81
Sensor, acoustic beam, 68
  back pressure, 53, 54
  bubbler tube, 55
  converging jet, 56
  diverging jet, 56, 57
  interruptable jet, 54, 55, 66, 67
  limit valves, 65, 66
  long range diverging jet, 67, 68
  vortex proximity, 57, 58
  vortex rate, 58, 59
Sensors, definition of, 162
  measurement of dynamics of, 98, 99
Shinn, J. N., 20
Signal generator, pneumatic, 93, 101
Signal to noise ratio, definition of, 84
Simulation, analog, 23
Small-signal analysis, vacuum triode, 6
Sorensen, P. H., 19
Spectrum, frequency, 1, 2
Stability criteria, 19, 22, 26
Staging techniques, 20
Static analysis, 20, 23, 69
Static characteristics, amplifiers, 176–180
  laminar flow restrictors, 181
  servo actuator, 180
  transistors, 12
  vortex rate sensor, 176, 177
Subscripts, definition of, 164
Switching characteristics, digital amplifiers, 74

measurement of, 97
Symbols, fluidic devices, 166–174
 functional, 164
 operating principles, 164
Symposium on Fluid Jet Control Devices, ASME, 16
System analysis, 13, 14, 18, 23
 nonlinear, 13
System design, 154–156
 analog, 25
 methods, 23
System requirements, 32
 definition of, 31
System synthesis, 18, 21, 23, 25

Test circuit, amplifier dynamic characteristics, 92
 amplifier static characteristics, 86
 element static characteristics, 91
 sensor static characteristics, 89
Testing technique, random, 26
Third Fluid Amplification Symposium, 22
Time delay, calculation of, 148, 149
 definition of, 83
 measurement of, 95, 96
 in proportional amplifier, 20
Transducer, definition of, 162
Transfer admittance matrices, 26
Transfer characteristics, analog amplifier, 71, 72
 definition of, 69
 graphical, 17
Transfer curve, 40, 120, 121
 calculation of, 197
 vacuum triode, 4
Transfer function, 40
 derivation of, 199–205
 numerical equivalent of, 150, 213–216
Transistor, characteristic curves, 12
 equivalent circuit, 12
 linear circuit, 12, 13
 models, 11, 12
 parameters, 11, 12
Transmission line theory, 8
Transmission of oscillatory pressures, 14
Transport time, 20
Tsui, K. C., 18, 22
Turbulence amplifier, 61–63
 definition of, 158
Two-port networks, 15
 transistor, 11

Vacuum triode, 4
Van Koevering, 17
Voltage analog, 22
Vortex amplifier, 48–50
 definition of, 158
Vortex proximity sensor, 57, 58
Vortex rate sensor, 58, 59
 equivalent circuit, 140, 141
Vortex valve, 19

Wall attachment amplifier, 59–61
 definition of, 158
 equivalent circuit, 139, 140
Wambsganns, M. W., 16
Warren, R. W., 15–17, 19
Wood, O. L., 20
Wright, C. P., 19

Yaw damper, fluidic, 25

STRATHCLYDE UNIVERSITY LIBRARY

30125 00190852 3

**Books are to be returned on or before
the last date below**